SLOW LIGHT

Invisibility, Teleportation,
and Other Mysteries of Light

SLOW LIGHT

Invisibility, Teleportation, and Other Mysteries of Light

Sidney Perkowitz

Imperial College Press

ICP

Published by

Imperial College Press
57 Shelton Street
Covent Garden
London WC2H 9HE

Distributed by

World Scientific Publishing Co. Pte. Ltd.
5 Toh Tuck Link, Singapore 596224
USA office: 27 Warren Street, Suite 401-402, Hackensack, NJ 07601
UK office: 57 Shelton Street, Covent Garden, London WC2H 9HE

British Library Cataloguing-in-Publication Data
A catalogue record for this book is available from the British Library.

SLOW LIGHT
Invisibility, Teleportation and Other Mysteries of Light

ISBN-13 978-1-84816-752-0 (pbk)
ISBN-10 1-84816-752-0 (pbk)

Typeset by Stallion Press
Email: enquiries@stallionpress.com

Printed in Singapore by World Scientific Printers.

To Sandy, Mike, Erica, and Nora
With love, as always

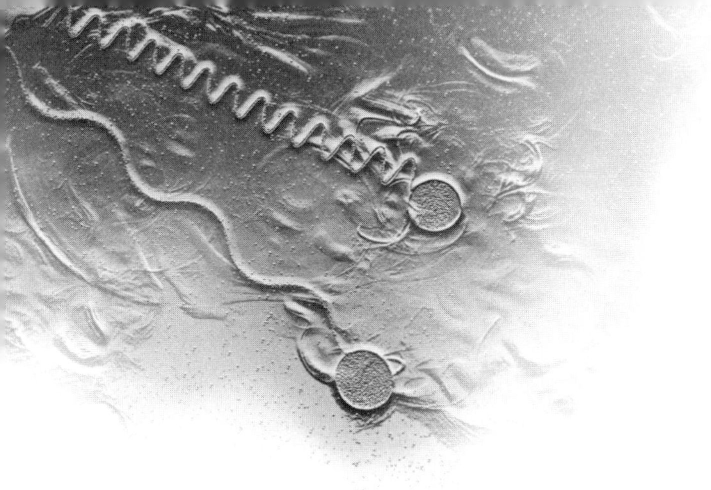

Contents

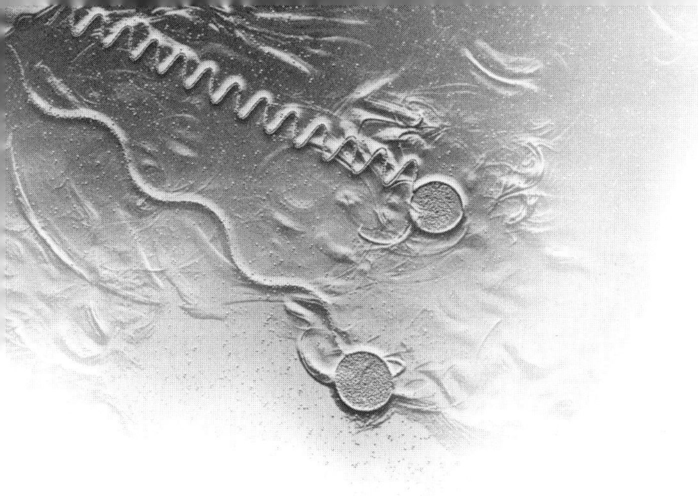

Introduction

Light is common, yet also wonderfully mysterious. After centuries of study, we still do not grasp how it functions as both electromagnetic wave and quantum particle, in the form of the photon; why it displays weird quantum effects; why, as Einstein's relativity tells us, its speed is the universal speed limit, and more.

But surprisingly, these puzzles have in the last twenty years engendered astonishing science fiction-like technology: quantum teleportation, where so-called "entangled" photons send their properties through empty space by unknown means; ways to slow light to a dead stop and also to speed it up; manipulation of light to make things and people invisible; atom-size sources that emit single photons, and stadium-size lasers meant to induce thermonuclear reactions.

Slow Light presents contemporary light research in popular form for those who wonder about things like whether teleportation is real (it is); how close we are to making a Harry Potter Invisibility Cloak (not for a while, but there's hope); whether lasers can produce hydrogen fusion (the jury's still out); and whether we can ever travel faster than light (unlikely). Along the way, I illuminate the science by comparing it to what science fiction and fantasy say about light.

Chapter 1 briefly surveys where light came from, what we know about it, and the technology that has arisen from its amazing properties, even those we don't understand. Chapters 2 through 6 go more deeply into the science and technology, area by area, in accessible form. Chapter 7 takes on the risky but exciting task of forecasting where this technology will go compared to what science fiction predicts.

If you like science, or science fiction and fantasy, or all of that, I think you'll enjoy riding along with me on a light wave — or a photon — and seeing where we end up.

Sidney Perkowitz
Atlanta, Georgia
December, 2010

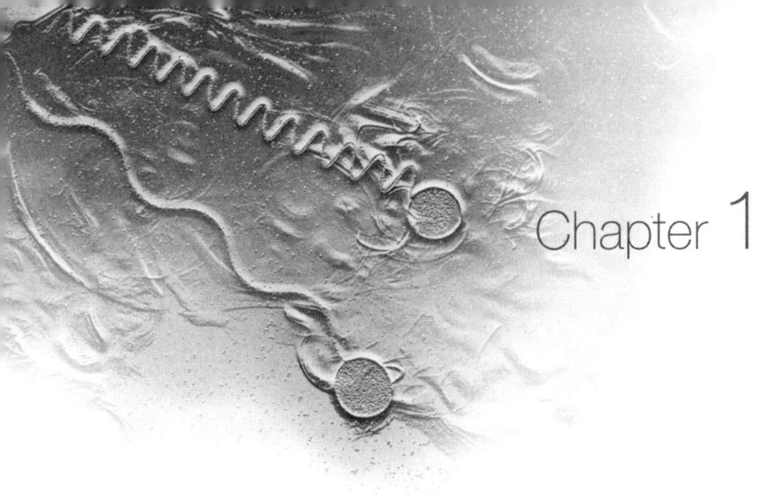

Chapter 1

What is Light? The Mystery Continues

You're reading these words under natural or artificial light, light that seems so ordinary that you take it for granted. Yet although light surrounds us, it is anything but ordinary. Its ancient role in the universe and its unique properties make it remarkable. Equally remarkably, light's science-fictionish behavior — such as its dual nature as wave and quantum particle and its ability to be teleported — can be applied in daily use and in advanced technology. Light also plays a continuing role in culture and in fantasy, from ancient myths to contemporary science fiction. All this has roots in humanity's long study of light.

To fully appreciate light's universal nature and how we use it today, let your mental vision roam beyond your immediate surroundings. Visualize our planet, then look beyond it to the Moon, the Sun, and the other bodies in our solar system; then further out to the stars, to the galaxies, and finally to the entire universe.

Light exists everywhere you turn your attention in our imaginary tour, part of it visible to human eyes, much of it detectible only by instruments. Some originates from a hot object such as an incandescent bulb or a star. Some is reflected from things on our world or off it: the pages of this book, a cloud, the planet Venus. And some has existed almost since the universe began, for light derives directly from the energy that filled the universe at the Big Bang. That explosion of reality began the cosmos 14 billion years ago with incredibly high temperatures. Among other things, those early conditions generated photons, the packets of energy that constitute particles of light.

It's not just that light is an ancient part of the universe. Light was also involved in the beginnings of matter. In the universe today, matter comes as a hundred different atomic elements and in sizes ranging from subatomic particles to stars and galaxies. It also comes in a variety of forms: as plasmas, that is, hot gases inside stars; as cooler gases between stars and in planetary atmospheres; in liquid form like the water and hydrocarbons found on Earth and elsewhere in the solar system; and as solids, like the rock and metal that make up planets and asteroids.

Different threads in the growth of the universe contributed to this diversity, among them, the evolution of light. When the universe was only a fraction of a nanosecond old, photons formed, along with other elementary particles such as quarks. According to Einstein's theory of relativity, energy and matter can change into each other, and so when photons collided, they turned into elementary particles with mass; for instance, electrons and their anti-matter partners, positrons. This may seem surprising, for we perceive light as pure and ethereal, beautiful in its intangibility. Yet even weighty materials, even dense metals such as lead or gold, can ultimately trace their origins back to light. The elementary particles that were formed from energy eventually combined into atoms, which in turn made the variety of matter that we know.

However, not all the light from the Big Bang became matter. Some, called the cosmic background radiation, still exists and travels throughout the universe. It is invisible to the eye but can be detected in any direction that we point our instruments. The distribution and wavelengths of this light are strong evidence that the Big Bang occurred.

Long before we knew this much about light, natural philosophers theorized about its nature. Light has always been important to humanity, starting with its reappearance each dawn as the sun rose. Ancient societies recognized that significance in their religious observances and in their gods of light and of the sun, such as Ra and Aten of the ancient Egyptians and Helios of the ancient Greeks. Some early thinkers, too, were drawn to consider light's properties. The Greek philosophers, forerunners of modern scientists, observed light and drew conclusions about its nature, some correct and some incorrect.

Around 300 BCE, for instance, the Greek mathematician Euclid, the "father of geometry," treated light rays as moving in straight lines, which is correct except near very large bodies like stars — but that was not known until millennia later, when Einstein's relativity predicted the effect and measurements confirmed it. On the other hand, some Greek pre-scientists such as the philosopher Empedocles (most famous for his theory of the Four Elements: Earth, Air, Fire, and Water) seemed to believe that human vision is enabled when a "visual ray" emerges from the eye to reach out to an object. Now we know that light must enter the eye to activate vision, and that nothing is emitted from the eye as part of the visual process.

Later thinkers, such as Aristotle and the Arab optical scientist Ibn al-Haitham or Alhazen, corrected some of the earlier misapprehensions and extended our knowledge of light. By the 17th century, tools and ideas were in place to begin the quantitative scientific study of light. Isaac Newton, perhaps the greatest scientist who ever lived, did so in 1665 when he examined the spectrum of sunlight dispersed by a glass prism that he had bought at a fair.

Newton concluded that light is made of "corpuscles" or individual particles, although the modern idea of the photon was still centuries away. His view remained the definitive understanding of light until scientist and polymath Thomas Young conducted a series of experiments beginning around 1801. In the most crucial of these, the famous "double slit" experiment, he passed light from a single source through two slits and recombined the two beams on a screen. The result was a series of alternating bright and dark bands. These absolutely could not be explained by Newton's corpuscles, which could add to give a result that was brighter than from one beam, but could never subtract to give an absence of light. Both bright and dark regions could appear, however, if light came in waves.

Unlike particles, waves are extended in space with alternating peaks and troughs. Lay two waves atop each other so that peaks match peaks and troughs match troughs, and "constructive interference" yields a new wave stronger than either of the originals. But shift one wave so that its peaks line up with the valleys of the other and you have "destructive interference," where the waves cancel each other to leave no activity or intensity at all. These phenomena explained the light and dark regions in the double slit experiment, and so Young's results seemed to establish that light is made of waves, not particles — though unlike the rise and fall of water in an ocean wave, exactly what was undulating in a light wave was then unknown.

The answer came in the 1860s from the Scottish physicist James Clerk Maxwell. His brilliant mathematical analysis of electricity and magnetism, known as Maxwell's equations, showed that when an electrical charge changes its motion — that is, accelerates — it produces an electromagnetic wave that travels through space. The connection to light came when Maxwell calculated the speed at which the wave would propagate and found it equal to the known speed of light. That had been accurately measured by then as close to today's accepted value of very nearly 300,000 kilometers per second (km/sec) or 186,000 miles/sec, a number that comes up so often that it has its own special symbol "c," supposedly to represent "constant."

By the end of the 19th century, it was clear that light is an electromagnetic wave defined by its wavelength, the distance between two successive peaks; or equally its frequency, how many peaks you count per second as a light wave travels past you (if you know one, you know the other because their product is equal to c). But new experiments were casting doubt on the wave theory. Measurements on the light emitted by a hot object, and on the photoelectric effect — where light impinging on a metal plate releases electrons — produced results that disagreed with how waves should behave.

In the 1890s, the German physicist Max Planck found that the problems with light from a hot body could be resolved if the energies in the process were emitted in discrete units or "quanta." In 1905, Albert Einstein took the idea further when he explained the photoelectric effect by showing that light itself comes in discrete packets of energy. These light quanta were later named "photons." Planck and Einstein both won Nobel Prizes, in 1918 and 1921, respectively, for these insights that gave birth to quantum mechanics.

Like Dr. Frankenstein's creature, Einstein's creation was remarkable but troublesome to its creator and others. Though the photon explained the photoelectric effect, it was at odds with the wave theory of light, and Einstein himself was never fully comfortable with the quantum nature of the universe. Nevertheless, starting in the 1920s, other scientists successfully developed quantum theory and applied it to matter as well as light. Theorists also went on to examine the physics of photons and electrons to produce quantum electrodynamics (QED), the quantum theory of electromagnetism. This work led to the 1965 Nobel Prize in Physics for Richard Feynman, Julian Schwinger and Sin-Itiro Tomonaga.

Einstein also extended our knowledge of light in a different way, through his theories of special and general relativity from 1905 and 1915 respectively. Special relativity shows that the speed of light is absolute; that is, unlike anything else in the universe, an observer measures a light speed of 300,000 km/sec no matter in what direction or how fast the observer or light source is moving. This differs from ordinary life. For instance, the driver of a car traveling at 60 mph measures the speed of a second car, traveling alongside at 60 mph, as zero relative to himself; but as Einstein himself noted, an observer on a light wave measures the speed of a second adjacent light wave as 300,000 km/sec, not zero. Another startling conclusion from special relativity is that nothing material can go faster than light, the fastest thing in the universe.

General relativity brought further understanding of how light behaves. Along with special relativity, the theory gives spacetime — the three spatial dimensions integrated with the fourth dimension of time — a central physical role in the workings of the universe. Gravity is explained as a distortion in spacetime due to the presence of mass. That distortion affects light, so a light ray no longer necessarily travels in a straight line, but traverses a curved path called a geodesic near a large object like a star. In extreme cases, a region of spacetime can become so warped that it traps light, that is, the region becomes a black hole.

These results from the theory of relativity are far from obvious or intuitive, yet Einstein's theories correctly describe the behavior of light, of gravity, and of the universe. Similarly, the photon picture and QED are highly successful in predicting the behavior of light and its interaction with matter. There's no doubt that the quantum nature of light is real, and photons take their place among the other elementary particles like quarks and electrons that make up the universe. Yet light also displays characteristic wave features such as constructive and destructive interference, and much of its behavior can be described by the theory of electromagnetic waves.

Along with the unusual aspects of relativity, this dual wave–particle nature of light is hard to accept. No one has ever brought the paradoxical elements of the duality fully under one theoretical umbrella. It is startling, and humbling, that after brilliant and sustained efforts over centuries to grasp the essence of light, we still do not understand how it embodies these two utterly different pictures.

Fortunately, incomplete understanding is no barrier to technological application. Scientists know how to use each half of the wave–particle duality as appropriate, and also how to apply relativistic principles. Without knowledge of general relativity, the worldwide global positioning system would be inaccurate; without theories of photons and quantum mechanics in general, we would not have conceived and invented many devices central to our technological society, such as computer chips, light emitting diodes, and lasers.

Nevertheless, the wave–particle duality of light is an aspect of quantum weirdness — one of those quantum phenomena that has no counterpart in human experience and perceptions and so seems inexplicable. The Heisenberg Uncertainty Principle is another strange feature. This states that there are certain combinations of physical properties, for instance the position and momentum of a small particle, that can't be simultaneously measured to a desired accuracy because the measurement process changes the properties. Any attempt to determine the position of a photon or electron adds energy to the particle, which changes its momentum, thus foiling the measurement of the momentum at the same time. This limitation is effective only at the small scales where quantum physics applies, and so we never experience it in daily life.

There is also the matter of superposition. In classical physics, a particle like an electron or photon has a single specific value for any of its physical properties. For instance, to pick a property useful for photon applications, a photon's electric field can be oriented or "polarized" along either a vertical or horizontal direction. But in quantum physics, that electric field has a probability of being oriented in either direction. It is not definitively vertical or horizontal until its direction is actually measured. Before the measurement, the photon is said to be in a "superposition" of states, with its field simultaneously vertical and horizontal — which seems to echo a famous bit of advice ascribed to baseball player and philosopher Yogi Berra: "When you come to a fork in the road, take it." Physicists are not entirely happy with this baffling image, but they accept it because it turns out to correctly describe the world around us.

Even more perplexing than quantum uncertainty and superposition is the exceedingly odd phenomenon called entanglement, which Einstein himself called "spooky." Consider two photons that were once closely associated, but have since been physically separated by being sent down two different paths. Select one of the two and measure its properties. Apparently instantaneously, the other changes in response even if it is many kilometers away, as if knowledge of their physical states passes from one to the other by an unknown process. This is akin to the classic science fiction idea of teleportation, as when Scotty on the starship *Enterprise* beams people and things across intervening gulfs of space. But entanglement is a real, serious, and growing research topic, an idea that has moved out of fantasy into science and potentially into new quantum technologies of computing and telecommunication.

The connection between quantum entanglement and teleportation is just one link between science fiction and the science of light. The speed of light is another, for its role as the universal speed limit hugely affects science fiction. Distances in the universe are so enormous that even if a spacecraft could travel at 300,000 km/sec — infinitely faster than NASA rockets actually move — it would require years to millennia to reach the stars outside our solar system. But science fiction writers want their characters to travel freely between stars and through galaxies, so they invent fictional ways to travel FTL, that is, faster than light. That's the origin of *Star Trek's* "warp drive" and other similar imaginary spacecraft propulsion systems. The idea is that the spaceship engines somehow distort spacetime to create a "hyperspace" that allows speeds greater than *c*, or that constitutes a cosmic shortcut or "wormhole" through which the spaceship can effectively travel faster than light.

Though general relativity does in fact predict the possibility of drastically changed spacetime in black holes and wormholes, and though there is strong evidence that black holes exist, there isn't as yet any firm idea of how to build a warp drive. But that hasn't stopped the quest to find ways to exceed the speed of light, which raises questions full of scientific interest: Are there loopholes in relativity? Has the value of *c* ever changed in the history of the universe? If something could go faster than light, how would it behave?

What probably never could have been foreseen, except by a science fiction reader or writer, is that researchers could also have become fascinated by light that is not fast but slow, creeping along at a few kilometers per hour and even brought to a full stop. In 1999, physicist Lene Vestergaard Hau, working at Harvard and the Rowland Institute in Cambridge, Massachusetts, used a new state of matter called a Bose–Einstein condensate (its creation was the subject of a Nobel Prize in Physics in 2001) to bring light down to the speed of a bicycle and then later to a dead halt.

Startling as it is to visualize a light ray moving at a snail's pace, at least two science fiction writers considered the meaning of slow light long before it became a subject for research. John Stith's novel *Redshift Rendezvous* (1990) postulated a hyperspace in which light is slowed by a factor of 30,000 to 10 m/sec, and worked out in careful detail the implications for daily life. Much earlier, in his well-known short story "Light of Other Days" from 1966, the Irish science fiction writer Bob Shaw proposed a way to make light crawl far more slowly than a ponderous glacier, then went on to explore what this could mean in human terms. (Both efforts were nominated for prestigious science fiction writing awards).

We are nowhere near creating any sort of hyperspace in which light might slow down, and the slow light in "Light of Other Days" seemed like pure fantasy when the story was written. Now after Hau's breakthrough and the work of other scientists, slow light can be made on demand; but it might seem a mere lab curiosity and even regressive, for after all, light's high speed is important for its applications. Whether as radio and television waves, or as infrared pulses traveling through optical fiber networks, light rapidly carries voice, music, images and data throughout the world. Certain applications, such as the global positioning system that orients a person or vehicle on the earth's surface, would not work well if light were substantially slower than it is. But although light's high speed is valuable, slow and stopped light can also lead to new applications in computing and telecommunications.

That isn't the end, though, of the manipulation of light speed. The same techniques that make light slow down can be used to do two astounding things: speed up pulses of light beyond c, and make light run backward, in the sense that the peak of a pulse leaves a given medium *before* entering it. These results seem like science fiction, not to mention that they seem to violate the relativistic speed limit and the universal principle of causality, that causes precede effects. Fortunately, careful consideration shows that so-called fast light and backwards light don't damage any fundamental theories; but these effects do show that the difference between science fiction and the science of light isn't always easy to determine.

Another way that light enters fantasy is in the wide use of lasers in science fiction. The power of concentrated light was recognized long before lasers were invented. In what is probably more legend than truth, there is an unsubstantiated story that around the year 212 BCE, Archimedes used curved mirrors to focus sunlight and set afire the wooden ships of an invading Roman fleet. A modern replication of this event shows that even if the enemy ships had obligingly held still to give the sunbeam time to heat them, the weapon would not have worked. Nevertheless, it shows that the ancients were well aware of the damage that light could do.

Millennia later, the power of concentrated light took on a more modern appearance in H. G. Wells' story from 1898, *The War of the Worlds*. In it, invaders from Mars deploy laser-like, invisible, but destructive heat rays. Ever since, lasers have been the armament of choice in many science fiction sagas, from the ray guns wielded by Buck Rogers and Flash Gordon, to an anti-weapon ray generated by the robot Gort in the film *The Day the Earth Stood Still* (1951), to the gargantuan Death Star laser that obliterates an entire planet in the original *Star Wars* (1977).

These fictional devices foreshadowed their real world versions. The carbon dioxide or CO_2 laser, invented in 1964, produces an invisible infrared beam like Wells' heat ray (as I got to know first-hand, having once been scarred by a CO_2 laser in my own lab). Laser space weapons were part of the Strategic Defense Initiative (SDI, also known as Star Wars) proposed by President Ronald Reagan in 1983. The plan was to place powerful X-ray lasers in orbit around the earth to destroy incoming nuclear-tipped missiles. The scheme came to nothing because of problems in developing the lasers and because the fall of the Soviet Union eliminated a major threat, but the military continues to develop other laser weaponry.

Despite the emphasis on weapons, lasers can also do constructive things. Low to medium power lasers send information over optical fiber networks, read bar codes in retail transactions, and perform surgery and other medical tasks. At higher powers, lasers can potentially provide clean energy for general use, although so far, only in science fiction. In the films *Chain Reaction* (1996) and *Spider-Man 2* (2004), lasers initiate hydrogen fusion, the process where hydrogen nuclei merge into helium and release energy.

In the real world, hydrogen fusion has been pursued for 60 years as a clean, practical power source. One current effort employs lasers far more powerful than those in *Chain Reaction* and *Spider-Man 2*. It is underway at the world's most energetic laser installation, the National Ignition Facility at the Lawrence Livermore Laboratory in California. This huge, arena-size complex is designed to deliver brief pulses of ultraviolet light carrying hundreds of terawatts of power into a small capsule containing hydrogen, in the hope of initiating a fusion reaction.

At the opposite extreme are devices devoted to making light at the lowest possible levels, one or two photons at a time, to study quantum entanglement and provide photons for quantum communication and computing. Single photons can be produced from single atoms or from quantum dots. These are very small semiconductor structures typically a few nanometers in dimension, that is, no more than a few dozen atoms across — so small that their properties are defined by quantum physics. Taken together, the light from the immense National Ignition Facility and a single photon from a quantum dot represent "extreme light," from ultrahigh power to the lowest power level short of zero.

Another area where science fiction has led the way, and the science of light follows, is invisibility. The magical power to make things and people disappear has a long literary and cultural history. It is used to make a point in Plato's *Republic* and it appears in early Teutonic mythology. This particular magic has been popular in fantasy fiction too, and remains compelling in modern times: think of the Ring of Power in *The Lord of the Rings*, which confers invisibility, and the Invisibility Cloak in the Harry Potter stories.

A different approach that replaces magic with science is represented by the H. G. Wells story "The Invisible Man" (1897), which describes the benefits and drawbacks of invisibility and sketches out a scientific method for achieving it. Another early story, "The Shadow and the Flash" (1903) by Jack London, gives an alternate scientific basis for making a person invisible. Probably the most famous non-magical invisibility apparatus in science fiction is the "cloaking device" in the *Star Trek* series that can hide an entire spacecraft, apparently by bending light so that it curves around the spaceship.

Echoing these fictional approaches, scientists are finding real ways to manipulate light so as to make people and objects invisible. The oldest technique is camouflage, first widely used in World War I. With modern technology, it goes beyond the familiar desert or forest patterns on soldiers' uniforms to make a subject appear transparent as glass and hence invisible. A second approach, stealth technology that renders aircraft virtually invisible to radar, began in the 1970s. The latest method, called "cloaking" as in *Star Trek*, redirects light rays around an object so that the light appears undisturbed by anything real. This technique uses artificial "metamaterials" that give new ways to control light rays. Unlike technology that reflects quantum weirdness, these approaches to invisibility use classical electromagnetic wave theory and can be applied to other kinds of waves, including sound waves, water waves, and seismic waves from earthquakes.

Indeed, a tour of the science of light is a tour of physics and technology, from the quantum world to the macrocosmic universe, from highly abstract theory to everyday applications, from old established ideas like Maxwell's equations to the cutting edge of modern science — and beyond it into speculative regions. It's striking that although the technology of light is embedded in our daily lives, much of it — especially many breakthroughs of the last 10 or 20 years — arises from the counterintuitive theories of quantum physics and relativity. Here are five important conclusions of these theories that will appear as we examine light:

1. Relativity I: Light in vacuum is the fastest thing in the universe. Nothing with mass and no information transfer can exceed that speed.
2. Relativity II: The path that light traverses follows geodesics in spacetime, whose shape is determined by nearby mass or energy; or to put it another way, light is affected by gravity.
3. Quantum weirdness I: Light is dual in nature, acting as particle or wave depending on the experiment.
4. Quantum weirdness II: A photon exists in a superposition of states until a measurement is actually made.

5. Quantum weirdness III: Photons can be entangled, so that a change in one of an entangled pair is instantly reflected in the other no matter what the distance between them.

These ideas and their consequences lie outside direct, visceral human experience and so they carry an air of mystery and fantasy that makes them easy to incorporate into speculative fiction. And truly, the line between speculation and fiction on the one hand, and the real science of light on the other, is blurred: sometimes the science fiction leads directly to the science, and sometimes the real science develops so rapidly that it outstrips the fiction.

In the remainder of this book, we'll look at both the real and the fantastic to see how they intertwine, beginning with a fundamental question about light: why does it move so very fast?

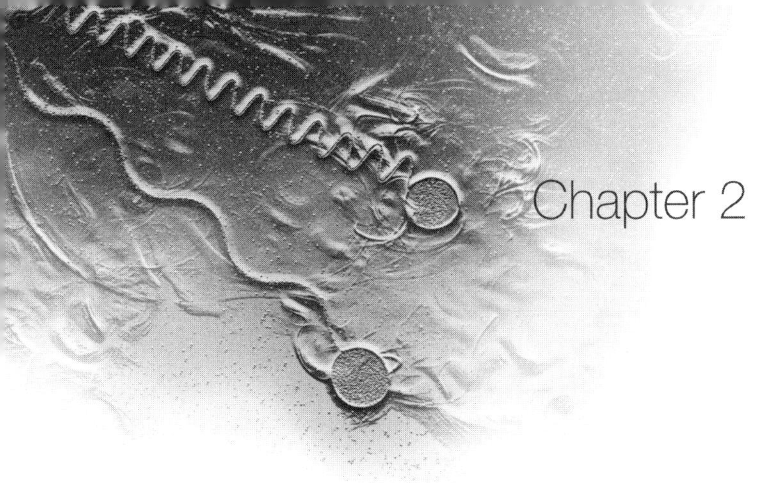

Chapter 2

Why is Light so Fast?

Along with Einstein's equation $E = mc^2$, probably the single best-known piece of popular physics lore also comes from Einstein's work; it is the fact that the speed of light is the speed limit for the universe. As a popular T-shirt puts it, alongside an image of Einstein togged out as a traffic cop, "186,000 miles per second is not just a good idea, it's the law."

That "traffic law" is well established now, but it has a long and tortuous history. Though from the earliest considerations of light it seemed apparent that its speed is very high and perhaps even infinite, centuries would go by before the speed was first found to be finite, and then measured. More time had to elapse before it was understood why light moves so quickly and why that is an ultimate limit.

Natural philosophers were interested in the speed of light long before theories of electromagnetism, photons, and relativity came along. The early Greek thinkers had no way to measure the speed but they still thought about it within their understanding of light. For example, in the first century CE the mathematician and engineer Hero (or Heron) of Alexandria concluded that light goes very fast. Using the incorrect idea of the visual ray that leaves the eye and reaches out to an object to enable vision, Hero reached a correct conclusion. He noted that one sees a distant object like a mountain or a star immediately upon opening the eyes. This must mean that light makes the round trip from eye to object and back very rapidly; hence, the speed of light is extremely high, perhaps infinite.

Not all the Greek thinkers agreed that the speed was infinite. Empedocles, who apparently believed in the visual ray, did not think so, but Aristotle did. The question remained open for centuries and generated varied responses. In 1637, the philosopher-scientist Rene Descartes analyzed how light must behave to make a lunar eclipse possible and concluded that light reaches the Earth from the sun "in an instant." The very next year, the great Renaissance scientist Galileo Galilei, best known for setting the Sun and not the Earth at the center of the solar system, put matters on a different track when he suggested actually measuring the speed.

In his *Two New Sciences,* Galileo noted first that "when we see a piece of artillery fired at great distance, the flash reaches our eyes without lapse of time; but the sound reaches the ear only after a noticeable interval." However, this only shows that light is faster than sound. To find its true speed, Galileo proposed using two experimenters with lanterns standing a known distance apart. One uncovers his lantern, and the other uncovers his when he sees the light from the first. With practice to eliminate reaction times, it would be possible, thought Galileo, to discern the time interval and determine whether light travels instantaneously. *Two New Sciences* suggests that this scheme was attempted for distances under a mile and that no time interval was detected, That's not surprising, since the travel time for that distance would be only a few microseconds. Galileo did however conclude that light travels at least ten times faster than sound.

Galileo's idea of measuring time of flight over a known distance was fundamentally correct, though to yield time intervals long enough to be measurable, it needed to be applied to large distances. That was finally done 38 years later in the city of Paris, when, as a memorial plaque at the exact site has it, "The Danish astronomer Olaus Römer (1644–1710) discovered the velocity of propagation of light at the Paris Observatory in 1676."

Like other great discoveries in science, Roemer's result (his name is commonly rendered as Ole Roemer) came because he seized on an unexpected observation. Roemer had not come to the recently built Paris Observatory (completed in 1672) to measure the speed of light, but to study the four largest moons of Jupiter. Galileo had discovered these through his brand-new telescope in 1610. If their orbital periods around Jupiter could be accurately found, they would serve as universal clocks that ships at sea could use to determine their longitude — an outstanding navigational problem of the time.

Roemer's data for the period of the moon Io as it orbited Jupiter showed a surprise. Instead of being constant, the period appeared to vary by up to 22 minutes when measured at different times of year. Various explanations were offered, but Roemer concluded that the discrepancy came about because it took time for light to travel from Io to Earth, the time varying with the distance between Jupiter and the Earth as they moved in their orbits.

This was the first clear-cut evidence that light has a finite velocity, with important consequences for understanding light and the entire universe. If light travelled at infinite speed, any images that astronomers gather even from the most distant galaxy would represent the galaxy at exactly that instant. This would have scientific value, but it would also deprive us of the ability to look back in time and observe the history of the universe. Travelling at finite speed, light that arrives from a cosmic object has taken time to reach us and so represents what the object was like that far back in the past.

The next step was to calculate the velocity. Christian Huygens, a leading optical scientist of the time, combined Roemer's time discrepancies with the distance that light traveled from Io to the Earth and obtained a value (in modern units) of 144,000 miles/sec or 232,000 km/sec. That's over 25% too low compared to the correct value because the distance was inaccurate, but it is of the correct large magnitude, and again, the significant fact is that it is a finite number.

Later, scientists found clever ways to measure the speed over earthly distances, as was first done by the French physicist Armand Hippolyte Louis Fizeau. In 1849, in a suburb of Paris, he set a light source and a mirror some eight kilometers apart and placed between them a rotating wheel with gear-like teeth. The wheel was used to measure the time of flight of a light ray from the source, which passed through the space between two teeth and was reflected back from the mirror. When the rotational speed of the wheel was set correctly, the returning ray intercepted the next gap rather than being blocked by a tooth, as could be observed by Fizeau. From the number of teeth and the rotation speed, Fizeau calculated the time of flight to obtain a speed of about 315,000 km/sec, only 5% different from today's accepted value for *c*.

Later, others measured *c* with greater accuracy. In 1862, the physicist Leon Foucault who had collaborated with Fizeau used a rotating mirror method to obtain a value of 298,000 km/sec, within 1% of today's accepted value. And in 1878, Albert Michelson, the first American to win a Nobel Prize in Physics, began highly refined measurements while a naval officer serving as an instructor at the U.S. Naval Academy in Annapolis. He continued measuring for many years and in 1926, published a value of 299,796 km/sec. This is very near the value 299,792.458 km/sec adopted in 1983 as definitive, based on contemporary highly accurate measurements by lasers and other means. This value enters into the modern redefinition of the meter, the basic unit of length, as the distance travelled by light in vacuum in 1//299,792,458 of a second.

The finite speed of light and its exact value were important for science and also soon entered general knowledge. In 1872, for instance, the French astronomer and science popularizer Camille Flammarion published *Lumen*, a mixture of science fiction and philosophical dialogue expressed through the character Lumen, a kind of disembodied spirit. Lumen points out how the finite speed of light enables us to see past events and calculates the time delays:

> If a volcano were to burst forth in eruption... we should not see the flames in the Moon till a second and a quarter had elapsed, if in Jupiter not till forty-two minutes.... a luminous ray coming from the star nearest to us, Alpha, in Centaurus, takes four years in coming.

These astronomical distances and times are still hugely important. But as light increasingly becomes part of modern nanotechnology, it's also useful to express c in units appropriate to small scales and brief time intervals. In those units, the speed of light is 30 cm/nsec, or almost exactly one foot per nanosecond.

A reliable result for c is also extremely important because it is a fundamental constant of nature, but the value of c has implications beyond simply knowing the number. For one thing, it was a major clue that light is electromagnetic in nature. The equations of electricity and magnetism that James Clerk Maxwell developed in the 1860s predicted that an accelerating electric charge would produce a traveling electromagnetic wave, as I wrote earlier, and also showed how to calculate the speed of the wave in space. When Maxwell did the calculation using known values for the electric and magnetic properties of space, it was, he wrote, "before I had any suspicion of the nearness between the two values of propagation of [electro]magnetic effects and that of light." To his surprise, the velocity he found for electromagnetic waves was, as he wrote in 1865:

> so nearly that of light, that it seems we have strong reasons to conclude that light itself... is an electromagnetic disturbance in the form of waves propagated through the electromagnetic field according to electromagnetic laws.

This theoretical insight was soon followed by experiments that confirmed the electromagnetic nature of light. Maxwell's result was the first scientific analysis of the nature of light since the days of the Greek philosophers that was supported by both physical theory and solid experimental results. Until the photon came along, it was the definitive understanding of light and its behavior.

The electromagnetic nature of light has important corollaries. Maxwell's equations apply to any medium, not only empty space, where a light wave's behavior depends on the medium's electric and magnetic properties. That explains why light slows somewhat in glass or water, and why it can be impeded more substantially to produce "slow light," and changed in other surprising ways in the right kind of medium, as I discuss later. It also implies that the speed of light in vacuum is higher than in anything denser. This is a hint that the speed of light in vacuum might be one of the higher velocities to be found in the universe, though that's not the same as saying that nothing can exceed that speed.

Rather, the speed of light as limiting value comes from Einstein's theory of relativity, but that is itself closely connected to light as electromagnetic wave. Einstein himself has written how, at age 16, he imagined himself riding on a light wave and examining an adjoining light wave, both moving at speed *c*. This would seem analogous to a commonplace situation such as sitting on a fast train going at 100 km/hr and looking over at another train on a parallel track going in the same direction at the same speed. An observer on the first train sees the second train as motionless, since it exactly keeps pace with him; or to put it another way, the speed of the second train relative to the first train is zero.

By analogy, as Einstein looked over at the second light wave, it should appear stationary. But Maxwell's equations don't allow a light wave with its peaks and troughs to halt and undulate in empty space; rather, like a shark, light must always keep moving, and at 300,000 km/sec. Instead of observing the second light wave as having a relative speed of zero, therefore, as would be the case for the trains, Einstein on his wave must still see it as moving at 300,000 km/sec.

This paradox and other considerations led Einstein in 1905 to his first theory of relativity, special relativity, which applies to objects moving in straight lines at fixed speeds. He built into the theory the axiom that any observer must measure the same constant speed of light *c*, regardless of the (constant) speed or direction of the observer or light source, and also the requirement that the laws of physics must be the same for any observer. From these he deduced some radical consequences. One is that lengths and time intervals depend on speed rather than having unchangeable universal values, so that the length of a moving body is decreased (known as length contraction), and time on a moving body runs more slowly (known as time dilation), as measured by a fixed observer.

Another radical result gave new meanings to the physical quantities mass and energy, including most famously the equation $E = mc^2$. This indicates that an object at rest has an energy E associated with its mass m, and that these two quantities can change into each other. Since the conversion factor c^2 is numerically large, this implies that a small amount of mass yields a large energy, and conversely that it takes much energy to produce a small mass — important factors in how the universe operates.

These results seem strange because we live at speeds much less than c where relativistic effects are too small to notice. But experiment after experiment shows that all the predictions of special relativity are correct. The theory has even changed the world; nuclear weapons and power plants work, for good or ill, because the relationship $E = mc^2$ is true.

The radical result from special relativity most relevant here is that nothing can go faster than light, as shown in the theory's mathematics and in its physical and conceptual consequences. Mathematically speaking, the equations of special relativity become problematic when evaluated for speeds greater than c, because then they require taking the square root of a negative number. The result is an imaginary number with no obvious physical meaning. Einstein alluded to this in his original paper that laid out special relativity when he wrote: "For velocities greater than that of light our deliberations become meaningless." (Actually, one interpretation of the imaginary result leads to the hypothetical faster-than-light-particles called tachyons, which appear in the next chapter; but at first glance, the imaginary result is the mathematical way of saying that the speed of light can't be exceeded. That's how Einstein took it, as would most physicists.)

The physical argument is one that Einstein himself made for material bodies. To get a body moving, that is, to accelerate it, takes a force — a push or pull. As the force is exerted, the body speeds up and gains energy of motion, called kinetic energy. To reach a certain speed, the accelerating force has to supply the corresponding energy. This depends on the body's mass, which resists acceleration through its inertia, and its speed. In relativity theory, the amount of energy needed to reach speeds near *c* increases sharply with every small step closer to *c*, so sharply that it would take infinite energy to actually attain light speed. This can't ever be supplied, or to put it another way, an object can never be pushed hard enough to attain the speed of light. As Einstein expressed it, "The velocity must always therefore remain less than *c*, however great may be the energies used to produce the acceleration."

We don't see anything like this relativistic behavior in ordinary life because fast but massive man-made objects like spacecraft require all the push rocket engineers can muster just to move at tiny fractions of *c*. However, subatomic particles like electrons have masses so small (the electron's is 9.1×10^{-31} kg) that they can be brought to speeds just short of *c* in particle accelerators, where the relativistic behavior of the kinetic energy is indeed observed.

A third argument is based on an assumption that seems inherent in how we understand the workings of the universe. It is the idea that causes produce effects, and in just that order. If event A is a cause and event B is its effect, then A must occur before B. This statement of the principle of causality may seem too ridiculously obvious to even mention, but in the world of relativity, causality is subject to revision. According to the mathematics of relativity, if it were possible to transmit information faster than light, it would also be possible to arrange things so that whereas A precedes B to some observers, B precedes A to others — which amounts to time travel. This is so upsetting to the order of the universe that we conclude that in addition to material things, information can never be transmitted faster than *c*. But as we'll see in the next chapters, it isn't always obvious what it is that carries information in a light wave.

These limitations raise a question: if objects can't reach the speed of light, how can the photons that make up light — which are one type of elementary particle — move at that speed? Part of the answer is purely mathematical. According to the equations of relativity that Einstein derived to link energy, mass, momentum, and velocity, any object with exactly zero mass automatically moves at light speed. It exists only at this speed, not needing a push of any sort to achieve it, and is also incapable of stopping.

If photons truly have zero mass, they fall in the category of these eternally moving particles, with properties consistent with relativity as well as Maxwell's equations; namely, they travel in vacuum at 300,000 km/sec and can never appear at rest relative to any observer. But this raises new questions: do we know that the photon's mass is really zero, and why does it take on that particular value?

Showing that a quantity is exactly zero is like the impossible task of proving a negative, but the evidence is strong that photons have zero mass. General considerations in quantum theory predict that this must be true, and quantum electrodynamics or QED, the quantum theory of the electromagnetic field, would not work properly otherwise. Since the results that come out of QED are among the most accurate predictions in all of physics, a failure of QED is extremely unlikely. Also a finite mass would change fundamental parts of electromagnetism such as how electrical charges interact and how magnetic fields behave. Measurements seeking these and other effects have succeeded in showing that the mass of the photon is at most 10^{-24} that of an electron. This is so tiny a quantity that the photon mass is universally accepted as exactly zero, which is why it travels at light speed.

Although the analogy is loose and imperfect, this makes some sense, since a massless object lacks inertia and so can effortlessly attain high speed. Science fiction writers have imagined ultrafast "inertialess" spacecraft, as I'll discuss later; but we don't encounter any such objects in the real world, so what is it that's special about photons? Actually, they are not entirely special, because they're not the only particles associated with zero mass. For a long time, it was thought that another elementary particle, the neutrino, is massless. There are three types of neutrinos, generated in radioactive decay and certain nuclear reactions. They're copiously produced by fusion processes in our sun and travel through space to the Earth, but interact only minimally with ordinary matter and so are hard to study. The latest results, however, show that neutrino masses are somewhere between 10^{-7} and 10^{-6} that of an electron — very small (though huge compared to the upper limit for the photon mass) but not zero, and so they travel nearly at, but not at, the speed of light.

There are still other particles, though, known or thought to be massless like photons and with similar functions, which suggests broader meanings for zero mass and the speed of light. The connection between photons and other elementary particles comes because photons are associated with electromagnetic forces, one of the four great forces of nature, which hold together atoms and molecules. The other three are the strong force, which binds together protons and neutrons into atomic nuclei; the weak force, responsible for certain other nuclear processes such as radioactive decay; and gravity, which holds together matter at the biggest scales.

In quantum theory, a force is carried or "mediated" by specific particles. Photons are exchanged among electrically charged particles, like snowballs tossed back and forth in a child's melee to cause a continual series of impacts, and produce electromagnetic forces. Similarly, particles called W (for weak) and Z bosons are exchanged within nuclei to create the weak force; and particles called gluons are exchanged between quarks, the elementary particles that make up protons and neutrons, and lead to the strong nuclear force.

Gravity is more complicated since it's explained by Einstein's general relativity, which is not a quantum theory but a geometric one. If gravity could be successfully brought into the quantum framework — which has so far proven extremely difficult — it would be mediated by hypothetical particles called gravitons. These may be found as scientists seek gravitational waves, the ripples in spacetime that propagate gravity according to the theory of general relativity. However, these waves are hard to detect and so gravitons may remain hypothetical for a long time.

In theory, gluons have zero mass like photons, and measurements indeed confirm that their mass is very small. In principle gluons travel at light speed, though normally they're not free to travel. Gravitons would propagate gravity at light speed and so would also have to be massless. But W and Z bosons break the pattern. They have substantial masses, nearly a hundred times that of a proton. So among the family of particles that mediate forces, photons, gluons and potentially gravitons are massless, whereas the W and Z particles are far from that.

This difference is puzzling and is part of a greater mystery, why the different elementary particles have the masses that they do. We understand a great deal about them through the quantum theory of elementary particles called the Standard Model — a major achievement that has been developed over decades from various Nobel Prize winning theories and experiments. But though physicists would like the Standard Model to be a "Theory of Everything," it is so far only a "Theory of Nearly Everything." It covers the particles and interactions that cause three of the four fundamental forces, namely, the strong, the weak, and the electromagnetic; but for the reason I mentioned earlier, it does not mesh with general relativity to explain the fourth force, gravity, nor does it explain the masses of all the particles.

Possibly the discovery of a particle called the Higgs boson will answer the questions about particle masses. The Higgs boson is supposed to set up a field through which the other elementary particles swim with varying degrees of resistance, thus accounting for their different masses. Finding this particular particle is a goal of the Large Hadron Collider near Geneva, Switzerland, the world's biggest particle accelerator at 27 km in circumference. After initial testing that began in late 2009, physicists plan to generate exceptionally high energies with the machine in the hope of yielding Higgs bosons.

Meanwhile, we can at least answer the question "Why is light so fast?" with the response "Because the photon is massless," although we don't yet understand why that's true. However, photons did provide an early entry into the whole issue of elementary particle masses. The first type of elementary particle to be explicitly identified was the electron, in 1897; yet when James Clerk Maxwell derived the speed of light decades earlier in the 1860s, he was for the first time finding the speed of a massless elementary particle, though without knowing it. It may be that *c* has its deepest meaning as the speed of any massless particle and as a fundamental property of spacetime; it became identified first as the speed of light only because light was accessible and could be analyzed long before gluons and gravitons were even imagined.

The speed of light became especially important after special relativity gave it a central conceptual role and made it the universal speed limit. Ten years later, Einstein worked out general relativity, his theory of gravitation that replaced Newton's gravitational theory from 1687. Newton's accomplishment was brilliant, but it assumed that gravitational interactions among bodies travel at infinite speed. Since this violates the universal speed limit, Newton's result is strictly correct only for velocities much less than c and Einstein had to derive a new relativistic theory. General relativity doesn't change the fact that c is the ultimate speed; but one of its intriguing, almost paradoxical outcomes is that it suggests ways to defeat that limit by distorting spacetime, the four dimensional vision of the universe that took on physical reality within general relativity.

These ideas and other ways to travel faster than light are the subjects of the next chapter, where first we'll see that though the speed of light seems unimaginably fast to us humans, in another sense it is not nearly fast enough.

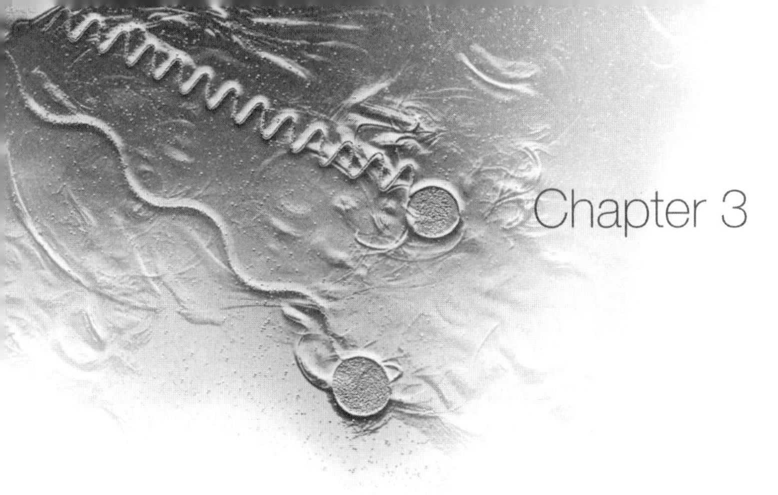

Chapter 3

Can Anything Go Even Faster?

We live in a fascinating universe, one that we yearn to examine by all means possible, starting with observation through telescopes and ending with human exploration. But in laying out the universe, Nature, or God — whichever you prefer — has played a huge practical joke that blocks our desire to learn more in person, for the expanses to be covered are almost beyond comprehension even at the speed of light.

The distances between various bodies in our solar system are already enormous. On our own planet, a trip around the equator is a sizable trek, 40,000 km; but that's insignificant compared to a voyage to the sun, 150 million km distant, or the outlying dwarf planet Pluto, which lies billions of kilometers from us. Yet these "neighborhood" solar system distances are in turn insignificant compared to distances between stars. The next closest star to us, Proxima Centauri, lies 4.2 light years (ly) away or about 40 trillion km (a light year is the distance light travels in one year at 300,000 km/sec, 9.5 trillion km) which is 270,000 times further than our sun. Other interesting parts of the universe lie even further. For instance, the center of our own Milky Way galaxy, where a black hole is thought to reside, is 28,000 ly or 2.7×10^{17} km distant.

Much as we would like to send astronauts to the stars, our technology is not remotely up to the task of covering these distances in reasonable times. The fastest vehicles humanity has ever made are spacecraft, and the record holder among them is NASA's New Horizons probe that headed toward Pluto in 2006 at a launch speed of 57,600 km/hr or 16 km/sec (the fastest manmade object ever is a spacecraft called Helios 2. In the 1970s, it attained a maximum speed of 250,000 km/hr as it orbited the Sun. However, this resulted from gravitational forces, not from the craft's own engine). This is many times faster than a commercial jetliner or even the speediest military fighter aircraft, yet not nearly fast enough to allow travel between stars in workable times. Even if a spacecraft could maintain that speed for the entire voyage, it would take nearly 80,000 years to reach Proxima Centauri.

If the spacecraft could move near or at light speed, thousands of times faster than the New Horizons probe, the situation would improve considerably, offering a sliver of hope that we might someday visit distant parts of the cosmos. At the speed of light, the Sun is reachable in about 8 minutes (that is, the Sun is 8 light minutes away), Pluto in around five hours, and Proxima Centauri in 4.2 years. A four year voyage is long but would probably be bearable. In the days of wooden sailing ships, crews endured years-long voyages, though they were able to break the journey with landfalls, and were not cooped up in the sealed environment of a spaceship. Today, however, the crews of nuclear submarines live in similar sealed environments for months at a time, showing that this kind of confined travel is possible (one difference though is that the spaceship crew would have to cope with the effects of time dilation).

But this comparatively rosy picture ignores the fact that at least for now, we can't accelerate anything the size of a spacecraft to even a reasonable fraction of the speed of light because the energy it would take is beyond our engineering ability. At 16 km/sec, that record-setting New Horizons probe was moving at only 0.005% of light speed. Nevertheless, it would open up possibilities if there were at least a theoretical way to match or exceed the speed of light, even if it can't yet be applied to spacecraft. Certainly science fiction writers need a way to move spacecraft and protagonists around the universe quickly enough to support real-time action. This science fiction tie-in is one reason that the science and the science fiction of faster than light (FTL) travel are so closely entwined.

At first glance, the problem doesn't seem insurmountable. There appear to be many real examples of "superluminal" behavior, where *something* — be it material object, information, or light itself — moves faster than *c*. Some of these situations don't even seem difficult to realize. Consider, for example, a theater marquee decorated with lamps that light up in sequence, or the illuminated news crawls and animated signs seen in New York's Times Square. In these, you can watch what looks like motion as lamp A turns on, then off, followed by adjacent lamp B doing the same, and so on. It's simple to calculate that with the right spacing and time interval between lamps, that apparent motion from A to B could happen at a speed greater than *c*. But nothing really moves from A to B. Neither matter nor information is transferred between the two points and the speed limit is not violated. The same objection applies to similar schemes, such as waving a laser back and forth on Earth, and projecting its beam onto the Moon's surface where the distance per time the beam covers expands enough to exceed *c*.

Light waves themselves, however, have features that really do move faster than *c*. There are also situations where material objects exceed the speed of light, but not in ways that violate the relativistic speed limit. Typically these behaviors occur when light travels through a medium, where its speed is different than in empty space. For example, the speed of light in water is only 225,000 km/sec. Electrons can easily be accelerated to higher speeds than that, and so electrons can travel through water faster than light (and in so doing, give off a characteristic blue glow called Cerenkov radiation). But Einstein's prohibition is only against exceeding the speed of light in vacuum, so nothing is violated here.

Another source of confusion lies in defining the speed of light. Finding the speed of a chunk of matter is straightforward, but a wave is extended in time and space, which makes its velocity harder to define. Several different velocities are associated with waves of any kind including light waves; phase velocity, group velocity, and signal velocity.

Phase velocity is connected to an idealized situation, a single monochromatic light wave with a particular frequency moving through empty space. A snapshot of it at any instant would show a sine curve with a definite wavelength extending to infinity in both directions. Any point chosen along its regular array of peaks and troughs, say at one of the peaks, represents the phase of the wave at that point. If you place a dot of red paint at that chosen phase, you can see the dot move through space at speed c, the phase velocity of that single light wave.

Surprisingly, if an electromagnetic wave is moving through something other than free space, its phase velocity can exceed c. That can happen if the wave is confined to a hollow metal channel called a waveguide, used in microwave technology; or if it passes through a plasma, a hot gas in which the atoms have become ionized into positively and negatively charged particles (plasmas can be found inside our sun and in the Earth's atmosphere).

A phase velocity greater than c seems to violate relativity, but it doesn't, because for electromagnetic waves it does not represent the transfer of matter. What may be more surprising is that neither is it the speed of transfer of information, called the signal velocity. That's because phase velocity is defined for an infinite wave that never changes, whereas conveying information requires change. That can be as simple as "on" to "off" or vice versa, like the binary "ones" and "zeroes" coursing through a computer; or as complex as the varying peaks and valleys of intricate sound waves. A stream of identical peaks and valleys in a monochromatic wave carries no information at all. Phase velocity is not signal velocity, and the relativistic speed limit doesn't apply to it.

Group velocity is more complicated. It arises when two or more light waves of different frequencies add up to make a wave shape or "envelope" other than a simple sine curve. The envelope has its own speed, the group velocity, determined by how the individual phase velocities of its component waves add up. In empty space, each wave has the same phase velocity c because that doesn't depend on frequency. The waves move in perfect step and the envelope also travels at speed c, so in vacuum, group velocity equals phase velocity c.

But in a medium, the phase velocity can depend on the frequency of the light, sometimes very strongly. This is called dispersion, and it is the reason that Isaac Newton could begin his seminal studies of light; dispersion is what makes a glass prism break white light up into colors. When dispersion is present, each component of the wave envelope moves at a different speed. In general, this changes the shape of the envelope as it traverses the medium, which it does at a group velocity — a complicated resultant of the individual phase velocities that can be very different from the phase velocity of any individual wave.

To see phase and group velocities in action, toss a rock into a pond and closely observe the resulting disturbance as it radiates out. It moves at a fast clip, but if you look quickly, you'll see individual waves within the disturbance moving toward its front, where they disappear. They're traveling at the phase velocity of water waves. For water that is not too shallow, that turns out to be twice the group velocity, which is the speed of the disturbance itself. Another example comes from Nick Herbert's book *Faster Than Light*. Imagine an inchworm making its way along the ground by forming its body into an arch and then flattening it. The arch travels along the inchworm's body at a certain speed, the phase velocity. Meanwhile the whole inchworm moves at a different speed, the group velocity.

Group velocity arises in experiments with light, because these frequently use lasers that produce brief pulses, which are described by group rather than phase velocities. Depending on the dispersive properties of the medium, such a group velocity can be much less than c. It can also be greater than c and even infinite! But this superluminal behavior does not violate relativity because although group velocity is meaningful, in this case it is not the signal velocity.

I'll discuss these matters further in the next chapter, which is devoted to slow and fast light. But for now, it's enough to focus on these two basic rules about the speed limit:

1. The limit is c, the speed of light in vacuum.
2. Anything moving faster than c does not violate relativity or causality unless it is matter or information.

Applying these rules eliminates trivial cases of apparent faster than light motion, but the universe can still surprise us with seeming examples of superluminal behavior. In the early 1970s, astronomers detected what looked to be speeds greater than c far out in space, when they observed the distant objects called quasars. These are extremely small and bright light sources, now thought to be black holes at the centers of young galaxies. The black holes generate enormous jets of plasma, and the measured speeds of the jets were as high as ten times the speed of light. If true, this would represent a massive violation of the relativistic speed limit.

Alas, these results are now accepted as due to incomplete knowledge. When an astronomer sees something moving far out in the universe, the motion looks transverse to his line of sight. The object very possibly also has a component of motion along the line of sight, but the astronomer can't measure that because he can't see into the third dimension. If the line-of-sight motion is toward the Earth, the object's apparent transverse travel time is less than its true travel time, because the light travels a shorter distance as the object approaches. An apparently shorter travel time mistakenly overestimates the transverse speed. If the object really is traveling near the speed of light — which these plasma jets are, up to $0.999c$ — the overestimate is enough to raise the perceived speed well above c (and if the object is headed away from the Earth, its speed is underestimated).

This mistaken analysis is emblematic of the reality: there is to date no observation of a meaningful transfer of matter or information at a speed greater than c. But this is discouraging for humanity's dreams of exploring the universe; and so, despite all evidence, scientists and science fiction writers have explored ways to sidestep the speed limit.

One possibility is that the theory of relativity is wrong. Like any scientific theory, it remains meaningful and useful only until new data or a new phenomenon arises that it can't explain, when a different theory is needed. But for now, there's little evidence that special relativity needs to be replaced. It is more than a century since Einstein proposed the theory. In that time, all its testable predictions — the constancy of c for observers and sources in different states of motion, length contraction, time dilation, increase of energy with motion, and the interconversion of mass and energy ($E = mc^2$) — have been thoroughly confirmed by measurement after measurement and by the use of relativistic ideas in real-life applications, in a great variety of situations.

Still, some challenges have been mounted against special relativity. Certain theories and some astronomical data suggest that the speed of light (along with other fundamental constants of nature) has not always been the same in the history of the universe. String theory, which some physicists think will resolve long-standing problems in our understanding of the universe, predicts that the speed of light is not constant. If either is true, that would strike at a cornerstone of relativity; but so far, the experimental evidence for these possibilities is scanty or nonexistent, so they can't yet be taken seriously.

Another point is that there is a certain apparent ambiguity about the speed of light in general relativity, Einstein's theory of gravitation, because the theory predicts that a light ray is curved in a gravitational field. This effect was used to confirm the theory in 1919, when data taken during a solar eclipse showed that our sun bent a light ray from a distant star by the small amount that Einstein had calculated. The curvature or bending implies that the velocity of light changes. As Einstein himself wrote:

> ... according to the general theory of relativity, the law of the constancy of the velocity of light in vacuo... cannot claim any unlimited

> validity. A curvature of rays of light can only take place when the velocity of propagation of light varies with position.

Einstein adds that this does not mean that general relativity overthrows special relativity, but only that the latter is a special case of the former when objects and observers are not accelerated.

In fact, the speed of light in vacuum never varies from *c*, including near a massive object; but the light beam curves because its travel time increases, changing its apparent speed. This is a result of gravitational time dilation, the prediction from general relativity that time moves more slowly the stronger the gravitational effect. In the 1960s, Irwin Shapiro of the MIT Lincoln Laboratory and coworkers exploited this outcome to test general relativity. They bounced a radar beam from Earth off the planets Mercury and Venus so that the beam passed near the Sun coming and going, and observed a 200 μsec time delay, as predicted by the theory.

Other potential problems with relativity arise from possible mismatches between it and quantum physics, as I mentioned earlier. Beginning with Einstein himself, who tried to develop what he called a unified field theory, theoretical physicists have spent years trying to mesh quantum physics, which explains the universe at the smallest scales, with general relativity, which explains gravitation and hence describes the universe at the largest scales. So far the two theories have not been successfully merged. Some scientists think that quantum weirdness, especially the spooky process of entanglement, may even eventually prove relativity wrong and upset the idea that *c* is a speed limit. But that possibility lies in the future, if it is even correct.

Given that special relativity is so well established, some physicists have worked within the theory to try to defeat the speed limit. Although relativity implies that it's impossible to accelerate an object to the speed of light, the theory may not disallow particles already moving at speed *c* or greater. In the 1960s, Olexa-Myron P. Bilaniuk of Swarthmore College and E. C. George Sudarshan at Syracuse University began considering how to fit what they called "metaparticles" with speeds greater than *c* into the relativistic scheme. The approach was extended in 1967 by Gerald Feinberg, of Rockefeller and Columbia Universities. In his theoretical paper "Possibility of Faster-Than-Light-Particles," Feinberg also introduced the wonderful name "tachyons" for these hypothetical particles, from the Greek word "tachys" meaning swift.

As I wrote earlier, when a speed greater than c is inserted into the relativistic equations, they yield imaginary numbers, that is, those involving the square root of -1. The mass of a particle traveling faster than c becomes imaginary, which seems to lack all physical meaning. But Feinberg reasoned that since a tachyon could never be at rest, its mass could not be measured anyway, and boldly assumed that tachyon mass could be taken as imaginary (or equivalently, that the mass squared is negative). He then showed that this led to acceptable non-imaginary values of energy and forged ahead to derive the unique properties of tachyons. Unlike conventional slower than light particles (called tardyons; particles that move exactly at the speed of light are known as luxons) that gain energy as their speed increases, tachyons always move faster than light and gain energy as their speed decreases. Both types approach infinite energy as speed approaches c, but tardyons approach it from below and tachyons approach it from above.

Despite some theoretical problems with these proposed new particles, including the general objection that they would violate causality and allow time travel, tachyons generated much interest. Over 600 papers about them appeared between 1962 and 1980, and some still do. Researchers analyzed tachyons theoretically and also sought them by appropriate experiments. Between 1990 and 2000, a flurry of results reported a negative mass squared for a particle called the electron neutrino. However, the question of neutrino mass is complicated and these experiments did not appear definitive in establishing that tachyons exist. At the moment, the general consensus is that tachyons have yet to be found, if they even exist, for further theoretical analysis suggests that they would be inherently unstable.

So although the idea of the tachyon is an instructive exercise in relativistic physics, like every other known approach it fails to break through the light barrier. However, if science is unconvinced about tachyons, science fiction embraces them. This illustrates how real science, at least at its more exotic or speculative limits, can be closely allied with fiction. Tachyons are a godsend to authors and filmmakers who need to send their characters around the universe at speeds that let exciting action unfold, rather than having to travel for centuries or millennia. It seems too that the very word "tachyon," because of its unusual Greek-origin spelling and engagingly catchy hard "ch" sound, lends a certain "science-ness" or science coolness to fiction.

Since tachyons first appeared in science fiction in the 1970s, they have been routinely invoked in works both classic and contemporary to enable FTL travel and communication as well as bridging past, present and future. In Isaac Asimov's award winning *Foundation's Edge* (1982), the fourth of his famous *Foundation* series, people can travel faster than light by being reduced to "incorporeal tachyons which no one has ever seen or detected." In Gregory Benford's *Timescape,* which also won major science fiction awards in 1980 and 1981, the physics of tachyons is discussed in detail and they are used to send messages back in time.

More recently, a villainous character in the popular science fiction-superhero story *Watchmen* (comic book series, 1986–1987; film version, 2009) makes strategic use of tachyons. He creates a shower of them to prevent Dr. Manhattan, a physicist who has gained extraordinary superhuman powers, from peering into the future, presumably because tachyons scramble cause and effect. Different television and film manifestations of the *Star Trek* universe refer to tachyons for varied purposes, including the cloaking of spacecraft, detection of cloaked craft that leave behind a tachyon residue, propulsion to FTL speeds, and penetration of spacecraft defensive shields. In two final examples, the alien character Prot in the film *K-Pax* (2001), discusses traveling at "tachyonic speeds" which are multiples of light speed; and the Asgard, the benevolent alien race in the *Stargate SG-1* television series (1997–2007) are also described as using tachyons for FTL space travel.

For all their popularity, tachyons are not the oldest or most widely used FTL device in science fiction. That honor goes to hyperspace, warp drives, wormholes, and all other means to somehow change or distort spacetime to change the speed of light, speed up a spacecraft, or take a short cut through the universe in order to travel faster than *c*. These have become so commonplace in science fiction that they're simply part of the generally accepted background, but it wasn't always so. They started with a scientific basis in the theory of relativity.

When special relativity was introduced in 1905, it set that speed limit of *c* for material objects, according to the argument put forth by Einstein himself, and for communication. But after general relativity came along in 1915 and its concepts began to be absorbed, somewhat paradoxically they seemed to provide loopholes to both scientists and science fiction writers.

Special relativity was startling among physical theories because it changed our concepts of space and time by showing that they vary according to the motion of an observer. The theory also showed that time is not independent of the three spatial dimensions of length, width and height as had been previously assumed. These ideas suggested that the right way to describe the universe is not by three space dimensions and a separate one of time, but as a true four dimensional continuum called spacetime, a view especially developed by the German mathematician Hermann Minkowski. He made the idea very clear in an address he gave in 1908 when he said: "Henceforth space by itself, and time by itself, are doomed to fade away into mere shadows, and only a kind of union of the two will preserve an independent reality."

General relativity, Einstein's theory of gravitation, deepens this approach by relating the real physical effects of matter and gravity to the shape of spacetime. In Newton's theory of gravity, a massive object like a star attracts a planet and holds it in orbit through "action at a distance," that is, a gravitational force that is exerted through space at infinite speed. But Einstein's theory does not use forces. Instead, the enormous mass of a star distorts the surrounding spacetime, which determines how other objects like planets move. The distinguished American theoretical physicist John Archibald Wheeler summed up general relativity in the phrase: "Spacetime tells matter how to move; matter tells spacetime how to curve."

To explain, first imagine a region of space where there are no stars and therefore no gravity. If you give an object like a planet a push to travel from one point to another, it moves in a straight line. In the terminology of general relativity it follows a straight geodesic, that is, the minimum distance between any two points is a straight line. Now introduce a star. Its mass "tells spacetime how to curve," meaning that it changes nearby spacetime so that the shortest distance between two points, the new geodesic, is now curved in three dimensional space.

If this is hard to visualize, consider our own planet. Absolutely the shortest distance between New York and Tokyo is along the straight line connecting them, but unfortunately that path bores through the Earth and so is unavailable to us, constrained as we are to the Earth's surface. Instead, the only option for least distance is the great circle route along the surface, which is the route that intercontinental aircraft fly; that is, the minimum-distance geodesics for travel on the Earth are curved, not straight.

Taking this as analogous to the spacetime surrounding a star in space, nearby objects like planets can move only along the curved spacetime geodesics that the star's mass induces, which is why planets follow orbits around a star. Light rays must also follow these geodesics, which accounts for the prediction of general relativity that a light beam is bent as it passes a massive object.

These notions show how general relativity gives warped spacetime (or just warped space, as it's often less accurately stated) physical significance including how it affects the behavior of light. As everyone knows by now, such effects become obvious and extreme inside a black hole (a name ascribed, by the way, to John Wheeler). That's the hugely distorted or warped spacetime surrounding highly dense matter, where gravity is so strong that once an object or even light enters, it cannot escape by any ordinary means. The idea goes back to 1783, but in 1916, the German physicist Karl Schwarzschild solved the equations of general relativity to derive a certain kind of black hole and put these entities on a firm theoretical footing.

Black holes lead to another result from general relativity, the wormhole (also named by Wheeler), a conduit between two different regions of the universe or even two different universes that traverses a "hyperspace" outside normal spacetime. The hyperspace tunnel might be only a kilometer long, whereas the regions it connected could be many light years apart. The result would be a stupendously efficient shortcut that allows a spacecraft to reach its destination faster than light traveling the normal route, like a traveler saving time by walking through a tunnel piercing a mountain while his companion walks the long way around the mountain. This form of FTL travel would be perfectly allowable because the spacecraft would never have to go faster than *c* within the wormhole. In the language of general relativity, the craft would cover distance faster than light "globally" without ever exceeding the speed of light "locally."

The first inkling of wormholes came from the Austrian physicist Ludwig Fromm, who in 1916 found that Einstein's equations permitted two Schwarzschild-like black holes to be linked by a conduit or throat. Later, in 1935, Einstein and his colleague Nathan Rosen determined that such a wormhole would collapse immediately after being formed. But in 1988, Caltech theoretical physicist Kip Thorne and his students found they could maintain a wormhole with "exotic material," that is, a negative mass (or equivalently, energy) whose gravitational effect would hold open the wormhole's throat. Thorne also went on to discuss time travel possibilities arising from such a "traversable" wormhole.

These ideas from general relativity have different levels of actuality. There is considerable evidence that black holes exist in nature. They can be formed when a massive star runs out of fuel and collapses under its own gravity. Then it dwindles down to a point of infinite mass density called a singularity, creating a black hole. There is also evidence that the centers of galaxies including our own Milky Way house black holes surrounding regions of very high mass density. But so far there are no indications that wormholes are formed naturally; no one knows where to find negative mass (though it is marginally less weird than the imaginary mass required for tachyons. Also its counterpart, negative energy, is not completely out of the question as I'll discuss later); and time travel remains controversial at best.

Regardless of reality or feasibility, these exotic but science-based ideas have been a rich source for science fiction, although that didn't happen immediately. Even in the 1920s, years after special and general relativity appeared, they apparently did not play a role in the pulp magazine science fiction that dominated the era. According to the *Encyclopedia of Science Fiction*, the science fiction writer John W. Campbell Jr., an ex-physics major, was probably the first to draw on relativistic ideas. He introduced warped space and the word "hyperspace" for FTL travel in his story "Islands of Space" from 1931. (This was among Campbell's earliest published works. He would go on to greatly influence science fiction as writer and then editor of the magazine *Astounding Science Fiction*.)

In the story, which is set in the future, Arcot, Wade, and Morey are three scientists who go off on interstellar adventures. They pay homage to special relativity, noting that "a man named Einstein said that the velocity of light was tops;" but also note that in general relativity, gravity and the behavior of light are intimately connected to warped space. This inspires the trio to build a "space strain" drive, which encloses their spaceship *Ancient Mariner* in a form of hyperspace that moves the craft at astonishing speeds, even when they test the spaceship just at half power. "We made good time!" says Morey "Twenty-nine light years in ten seconds!" But even 2.9 ly/sec is paltry compared to the speed at full power of 23 ly/sec, or better than 700 million times *c* — so fast that a spacecraft would reach our neighbor star Proxima Centauri in a fifth of a second rather than the 4.2 years it would take at the poky speed of light.

This performance must have impressed other science fiction writers, because warp drives, hyperspace, and so on have become science fiction staples in different ways. Sometimes FTL travel occurs within an entire hyperspace with unusual properties. For instance, in Isaac Asimov's *Foundation's Edge*, tachyons enable FTL travel within a hyperspace. In it, as one character puts it, the galaxy shrinks to a "nondimensional dot" where

> speed does not really have a meaning. Hyperspatially the value of all speed is zero and we do not move; with reference to space itself, speed is infinite. I can't explain things a bit more than that.

The reader might be excused if he or she is also baffled by a speed that's both zero and infinite, but John Stith's novel *Redshift Rendezvous* (1990) makes a simpler statement. In it, the spaceliner *Redshift* operates in a hyperspace where it moves at exactly 1,024 times the speed of light relative to normal spacetime (This hyperspace has another interesting property. Though it speeds up motion relative to the conventional universe, the speed of light in it is much lower than 300,000 km/sec, a phenomenon I'll explore in the next chapter).

Other science fiction authors followed Campbell's lead, surrounding their spacecraft with an artificial bubble of hyperspace that propels it through regular spacetime faster than light. That's the case for the best known science fiction FTL system, the "warp drive" used in the *Star Trek* universe. The drive generates a warped spacetime bubble that surrounds a spacecraft like the *Enterprise* and moves it much faster than light, yet also maintains the normal laws of physics and suspends time dilation within the ship. This doesn't take infinite energy but does require very high power, which comes from the energy released when matter and antimatter mutually annihilate each other in the *Enterprise's* engines. The speed of the spacecraft can be calculated from the warp factor; "warp 5" corresponds to 214 times the speed of light, and "warp 9" to more than 1,000c. (Note that neither the *Redshift* nor the *Enterprise* comes close to beating the speed Campbell gave his *Ancient Mariner* in 1931).

Rather than claiming "local" speeds in excess of c, other writers use warped space in the form of wormholes to provide cosmic shortcuts. That's what Carl Sagan did in his story *Contact* (published 1985; film version, 1997). In an example of science fiction influencing science, while writing the story Sagan sought a scientifically valid way to allow his hero, radio astronomer Ellie Arroway, to travel quickly through the universe. He asked theoretical physicist Kip Thorne for ideas, which led Thorne to think about wormholes. The result was that Thorne developed the concept of traversable wormholes and Sagan had a way for Arroway to cross the galaxy and reach the aliens whose radio messages she had detected.

There are other ideas for travel at or beyond the speed of light where boundaries between science and science fiction are blurred or have become less rigid than one might think, not all of which involve general relativity. An example is one approach suggested in science fiction, which is to eliminate inertia, the property of matter that resists motion and is expressed in mass. If that could be done, even ordinary matter would behave like a photon, whose zero mass allows it to move at light speed as dictated by special relativity.

Inertialess spaceships first showed up in E. E. Smith's ("Doc Smith") epic space operas from the 1930s and 1940s about the Triplanetary League and the Lensmen of the Galactic Patrol. Other science fiction writers later used the idea too. Smith's stories didn't mention relativity or its limits and his ships moved faster than light, reaching top speeds of 91 ly/sec in interstellar space, better than Campbell's *Ancient Mariner*. At the time Smith wrote and for long afterwards, there was no theoretical basis to manipulate the properties of mass as Smith imagined. But now, with the advent of the Higgs boson as perhaps the origin of mass, the idea of manipulating inertia at a fundamental level seems less like pure science fiction.

NASA has also seriously explored boundaries between science and science fiction in a quest for new ways to propel spacecraft. In 1996, recognizing the enormous difficulties of interstellar travel, NASA started its Breakthrough Propulsion Physics (BPP) program which examined radical ideas for spacecraft drives — ideas so radical that they would require significant breakthroughs in physics. One proposal was for a "diametric drive," which would create an asymmetric gravitational field that would continually propel a spacecraft in a given direction. Since gravitational interactions between two pieces of ordinary matter are symmetric (what the first mass does to the second, the second does to the first) this drive would require exotic matter in the form of negative mass — the same requirement Kip Thorne put forth for traversable wormholes — that could be arranged so that there is always a net force propelling the spacecraft.

Another idea considered by BPP generated excitement when broached because it is the closest we have come to a true warp drive. In 1994, theoretical physicist Miguel Alcubierre at the University of Wales published a paper with the intriguing title "The warp drive: hyper-fast travel within general relativity." The abstract was also fascinating, stating: "It is shown how, within the framework of general relativity and without the introduction of wormholes, it is possible to modify a spacetime in a way that allows a spaceship to travel with an arbitrarily large speed."

The basic idea behind this is easily expressed; by expanding spacetime, an object can be made to move away from a given point at any desired speed. This can be greater than c, because the speed of light is being exceeded only globally, not locally. Similarly, by contracting spacetime, an object can be made to approach a given point at any arbitrary speed. Using the equations of general relativity, Alcubierre combined these two ideas to calculate the overall shape or "metric" of a spacetime bubble that would contract in front of a spacecraft and expand behind it, thus propelling it in the direction of the contraction — the very model of a science fiction warp drive that even includes the desirable feature of eliminating troublesome time dilation effects.

But though the reasoning is convincing, engineering execution of the idea remains elusive, to say the least. That's because, like a traversable wormhole and as Alcubierre himself noted, this particular metric requires negative mass. The mass would also have to be huge because spacetime is not easily distorted. By one calculation, it would take a negative mass many orders of magnitude larger than that of the entire universe to move a spacecraft across the galaxy. So for the moment at least, the Alcubierre warp drive remains only a tantalizing mathematical exercise. Perhaps because it and other radical ideas were not producing tangible results, BPP was terminated in 2003 after over $1 million had been spent on it.

The closing of the BPP project is a metaphor for the status of FTL travel. As much as NASA and the scientific community, would-be galactic explorers, and science fiction fans would like to see some realistic possibility of FTL travel, at the moment there is not much hope for that. Einstein's special relativity remains solid, tachyons have not been found, and the end runs suggested by general relativity encounter practical difficulties. One small glimmer of possibility, however, is that quantum theory combined with gravity, or pure quantum weirdness, may show us ways to realize such seeming impossibilities as negative mass. I'll have more to say about this in a later chapter.

Nevertheless, this particular marriage of science fiction and science has forced us to dig more deeply into the properties of light and underlines its central role in the universe. And though we're foiled by the speed of light as an absolute limit, there is no limit to how slowly light can go, as science fiction writers have explored and scientists have learned. They've learned, too, that in a sense light can travel faster than light, though not in violation of Einstein's speed limit. All this I'll examine in the next chapter.

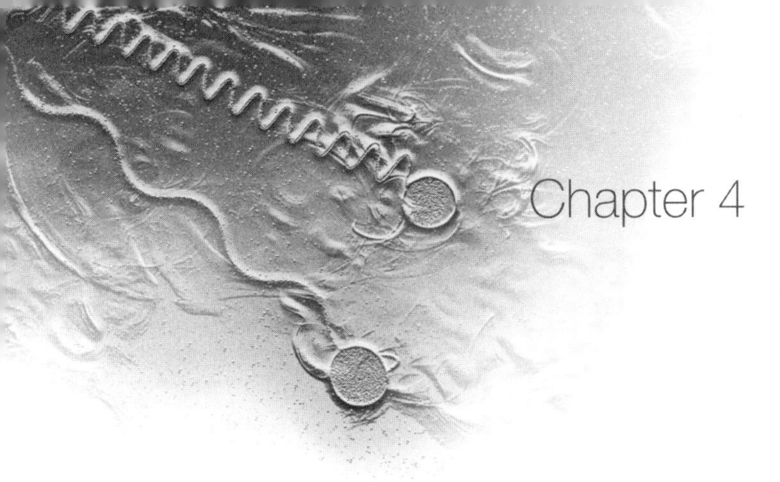

Chapter 4

Slow, Stopped, Fast, and Backwards Light

The great contemporary physicist Richard Feynman won the Nobel Prize in Physics in 1965 for his work on the highly abstract quantum theory of photons, but he is also credited with drawing attention to nanoscience, a field with widespread application. He did that in a famous address he made in 1959 called "There's Plenty of Room at the Bottom," referring to the fact that the science and technology at very small scales would be of immense fundamental interest and great practical importance — as of course has proven true in the active field of nanotechnology.

In a similar vein, though there's an undeniable attraction to imagining ways to move at or beyond the speed of light, there's plenty of fascinating science and useful technology at the low end of the speed spectrum, where light is brought to the pace of a bicycle or even to a standstill. More surprising yet, the same basic process that slows light can also be used to speed it up faster than c, and even to travel backwards in time — though, as in every other case, careful consideration shows that these results do not violate relativity and causality.

Maxwell's equations, Einstein's relativity, and numerous measurements all combine to establish that the speed of light is a universal constant of nature with the value of almost exactly 300,000 km/sec.

But the word "constant" is slightly misleading, because light travels at that speed only in empty space. The speed decreases in any ordinary medium through which light travels, very slightly in air; more substantially in water, glass or other transparent or semi-transparent materials, typically by a factor of two or less, depending on the material.

The lower speed of light in a medium can be explained in either of the two views of light. In the wave picture, the oscillating electric field in the light wave couples to electrically charged particles in the medium, usually electrons, and sets them vibrating. These moving charges create new electromagnetic waves that interfere with the original ones and result in waves that move more slowly. In the photon picture, incoming photons combine with quantum particles arising from the atoms and molecules in the material to form new types of mixed quantum particles. These have masses greater than zero, and so must move more slowly than 300,000 km/sec.

Whether using waves or photons, the standard way to describe the speed of light in a given medium is by its index of refraction, defined as the ratio of the speed c in empty space to the speed in the material and represented by the symbol n. Refraction is the process whereby a ray of light changes direction as it moves at an angle from one medium into another, resulting in the familiar phenomenon of the stick that is seemingly bent when partly placed in water. This happens because light waves move at different speeds in the two media. The ratio of the speeds, the refractive index, can be used to calculate the change in direction as light moves from medium to medium (refraction now seems so straightforward that it is taught in introductory physics courses but it took much time and effort to correctly derive Snell's Law, the equation that relates the change in direction to the value of n).

Since the speed drops in a medium, n is always greater than one for ordinary materials. For instance, $n = 1.33$ for water, meaning that the speed of light in water is 225,000 km/sec rather than 300,000 km/sec. Other typical values for n are 1.0003 for air, so the speed of light in air is barely different from that in vacuum; and in the range 1.5 to 2 for various kinds of plastics and glass, representing speeds of 150,000 to 200,000 km/sec. (As a sidelight, the value of n for diamond is unusually high at 2.4. This means that light is bent sharply as it makes the transition from air to diamond and vice versa, which is why a diamond artfully cut with facets at the correct angles sparkles so brilliantly).

So although it's not obvious, we are always surrounded by examples of light moving more slowly than the speed c. This is crucial for refraction to occur, but otherwise, the degree of slowing is not very significant in daily life. A speed of 150,000 to 200,000 km/sec in glass is still enormously fast. It means that a person watching an event through a window sees it only an imperceptibly small fraction of a nanosecond later than someone watching directly. It's true that data bits in the form of infrared pulses travel through the optical fiber network at the speed of light in glass, not vacuum, but that's still fast enough to travel around the Earth four times a second and faster by far than electrons in copper wire.

However, if the speed of light were innately much lower than this or could be artificially slowed by a large factor, that would affect both the natural world and our technology, from the Internet to the Global Positioning System. But although there's no theoretical reason why the speed of light can't be deliberately reduced by a large amount, this isn't easy to do. That's why it happened in fiction long before it was really accomplished. And even in fiction, slow light has been treated far less often than FTL travel, but some science fiction and fantasy writers have speculated about the possibility.

One example arises in Terry Pratchett's popular *Discworld* fantasy series, consisting to date of over three dozen novels. The first, *The Color of Magic,* was published in 1983. The series is set on a 10,000 mile wide disc that rests on the backs of four elephants, standing in turn on a gargantuan turtle that swims through space. As the first title of the series indicates, this imaginary world supports magic. One outcome is that light is slowed to the speed of sound, a few hundred miles an hour, which leads to such fascinating effects as light visibly creeping into a valley at dawn and draining away at sunset.

A different approach, science fiction rather than fantasy, is central to John Stith's *Redshift Rendezvous* (1990). This well-regarded story (it was a finalist for the prestigious Nebula award) deeply explores the idea of extremely slow light. Much of the action takes place aboard the space liner *Redshift,* which rapidly covers distances between stars by cruising in layer number ten of a multilayer hyperspace, where the ship's speed relative to ordinary space is effectively 1,024*c*. But the properties of spacetime in layer 10 are such that in the *Redshift* itself, light travels at only 10 m/sec (35 km/hr or 22 mph), 30 million times slower than in normal space, and comparable to the speed of a fast sprinter.

This leads to intriguing phenomena aboard ship. Some are direct consequences of the low speed of propagation for light: for instance, when a lamp is switched on, the light can be seen to bounce around a cabin until it reaches equilibrium; or a passenger can rotate rapidly enough in front of a mirror to see her back, since it takes a perceptible time for light reflected from her body to reach the mirror and return to her eyes.

The colors of things change too, because of the Doppler effect. That's the wave phenomenon that increases the pitch of a sound when its source and an observer approach each other, and decreases it when they move apart, as you can hear the next time a fire engine rushes by with siren screaming. The shift also applies to light waves, but is significant only when source or observer is moving at a large fraction of the speed of light, so we don't see Doppler shifted light in our daily activities. Aboard the *Redshift,* however, a person can run fast enough to see objects ahead tinged with blue and those behind tinged with red, corresponding to light Doppler-shifted up or down in frequency.

Similarly, passengers and crew on the *Redshift* experience effects predicted by special relativity that are too small to discern at typical speeds in the normal universe. Now these appear at easily attainable speeds, so a passenger running down a corridor sees its length contract. Other effects reflect predictions from general relativity about how light and time behave in a gravitational field. The *Redshift* generates its own internal gravity by a technology that puts the equivalent of a small planet at the core of the spaceship's spherical shape. The result is a gravitational field that gets weaker further from the core, like the Earth's. According to general relativity, the rate at which time flows depends on gravity, so a clock aboard ship keeps time differently near a person's feet and near his head. Another relativistic outcome is that light gains energy and therefore becomes higher in frequency or more blue as it plunges into a gravitational field, and loses energy as it climbs out, so becoming redder. A passenger on the *Redshift* sees these color changes depending on whether the light source is higher or lower than her eyes.

Also, as I discussed previously, gravity bends light, but it takes a truly massive object to produce much deviation because the light flashes by so quickly. At the low speed of 10 m/sec, though, the bending is significant in the time it takes light to cover just a few meters, so aboard the *Redshift*, a flashlight beam follows a curved rather than a straight path.

These bizarre effects are glimpses of what our world would be like if *c* were decreased to a very low value. In the story, the reduction in speed is made plausible by assuming that the *Redshift* operates in a hyperspace that differs from normal space. But as I wrote in my earlier discussion of spaceship drives, although extremely distorted spacetime exists in theory and can be found inside a black hole, we do not know how to produce and use such "hyperspaces" for our own purposes, let alone tailor one to radically slow down light.

There are however other ways to think about slowing light, as in Bob Shaw's science fiction short story from 1966 "Light of Other Days" (winner of a Nebula award) and its sequels. Shaw proposed a different method, which has some scientific justification, to slow light even more than in the hyperspace of *Redshift Rendezvous*. The idea is to use a medium with a huge index of refraction. This had earlier appeared in the short story "The Exalted" (1940) by the classic science fiction and fantasy writer L. Sprague de Camp. In the story, the scientist character Prof. Methuen finds a way to temporarily raise the index of refraction of a pure glass rod to "a remarkable degree... light striking the glass is so slowed up that it takes weeks to pass through it" (He also adds: "I can shoot an hour's accumulated light energy out the front end of the rod in a very small fraction of a second," which seems like a precursor of the laser).

Shaw's stories extend the idea by featuring "slow glass," which is transparent like ordinary glass, except that light traverses it even more slowly than in Methuen's rod. Once the light from a scene enters a thin sheet of the stuff, years elapse before it emerges from the opposite face. Thus the emerging light carries the past, just as the light from a star 1,000 light years away shows the star as it was 1,000 years ago.

This gives slow glass a wonderful property. As Shaw describes it:

> ... one could stand the glass beside, say, a woodland lake until the scene emerged, perhaps a year later. If the glass was then... installed in a dismal city flat, the flat would... appear to overlook the woodland lake... the water would ripple in sunlight, silent animals would come to drink, birds would cross the sky, night would follow day, season would follow season. Until one day, a year later, the beauty... would be exhausted and the familiar gray cityscape would reappear.

Slow glass carries emotional weight too, at least for the story's middle-aged character Hagan, who sells slow glass that holds scenic beauty gathered near his cottage in the Scottish Highlands. He is obsessively devoted to one particular pane of the material that shows his young wife and son, killed years before in a road accident. This sustains Hagan's sad illusion that they are still alive, but when the light has finally moved completely through the glass, he'll be utterly bereft.

Besides Hagan's poignant story, Shaw gives enough detail to make quantitative judgments about slow glass, since the story states that it takes ten years for light to work through a pane a quarter-inch or 6 mm thick. Compared to the ten light years or nearly 10^{14} km that light would cover in ten years of travel through free space, that internal speed is almost inexpressibly slow. It works out to 0.02 nm/sec, meaning that light crawls along, barely covering the diameter of a smallish atom like hydrogen or helium each second.

That speed corresponds to an index of refraction in the thousands of trillions, a value of *n* so high that no ordinary material could provide it. There are other problems as well with Shaw's conception of slow glass. Ten years worth of light reflected from a scene represents a lot of energy. It's questionable whether a thin pane of material could hold it. But there's an even more basic objection. The huge mismatch in speed between light outside the slow glass and that inside makes it difficult to transfer energy to the glass. With its high value of *n*, slow glass would reflect 100% of incoming light, so in principle no light would enter the material at all.

Still, in using the idea that light behaves differently in a material medium than in empty space, Shaw's approach is more believable than the hyperspace in *Redshift Rendezvous.* But since no ordinary material could work as Shaw envisions, when light was first really slowed to a speed at human scale, the feat required an exotic medium, a gas of atoms cooled nearly to absolute zero. And to reach the very lowest speed attained by Lene Hau, the physicist who slowed light by a factor of many millions, the gas had to become colder and more exotic yet, coalescing into a form called a Bose–Einstein condensate or BEC.

Scientists didn't start studying these cold atomic systems because of their unusual optical properties. Part of the motivation is learning how to reach extremely low temperatures near absolute zero, that is, a temperature of zero on the Kelvin scale (temperatures on this scale — named after the 19th century British physicist Lord Kelvin — are calculated by adding 273 to the temperature in degrees Celsius and are expressed as "kelvins" or by the symbol "K." Water boils at 373 K and freezes at 273 K; typical room temperature is 300 K. Conversely, 0 K, absolute zero, corresponds to — 273° Celsius). One of the three basic laws of thermodynamics says that absolute zero cannot ever be reached but only approached, so it's a challenge to physicists' ingenuity to find ways to climb down from 300 K to get as close as possible to 0 K.

An equally important reason to study ultracold systems is that they display fascinating quantum physics. At low temperatures, the thermal energy possessed by atoms and electrons is very low, and quantum behavior emerges that can't be seen at normal temperatures. For instance, helium gas condenses into liquid helium at around 4 K, and then cooled even further becomes a strange kind of quantum fluid that spontaneously crawls out of containers. Electrons in certain materials take on quantum properties at a few degrees above absolute zero that make them superconducting; that is, they carry electrical current with absolutely no losses to heat.

Go even colder, and it becomes possible to form a BEC, a group of subatomic or atomic particles that have collapsed into a single quantum entity. The theory behind this unusual state goes back to 1924 when the Indian physicist Satyendra Nath Bose considered how groups of photons behave. Photons belong to one of two general classes of elementary particles, called bosons, which can congregate at the same energy and share just one quantum state. Other elementary particles such as W and Z bosons, and certain atoms, also belong to this class. The other category, which includes electrons, is called fermions after physicist Enrico Fermi and behaves differently. Each fermion in a group must occupy a separate quantum state, with no common energy level. (The differences between bosons and fermions relate to their spin, which is the quantum mechanical analog of the particle visualized as a tiny ball rotating around its own axis. I'll say more about spin later).

Following Bose's work, Einstein showed that boson-like atoms could condense into a group all occupying the same quantum level and behaving like a single quantum particle, a kind of super atom. This would be of huge interest because the condensate would approach macroscopic size, unlike most other quantum systems. But there was a catch: the coming together would happen only at extremely low temperatures barely above absolute zero.

The confirmation of Bose–Einstein theory had to wait until physicists could reach those temperatures. At last, in 1995, Eric Cornell and Carl Wieman, working at the University of Colorado and two government laboratories in Boulder, Colorado, cooled a gas of rubidium atoms to millionths of kelvins and created a BEC containing some 2,000 atoms. Wieman described this sizable quantum system by saying: "We brought it to an almost human scale. We can poke it and prod it and look at this stuff in a way no one has been able to before." Cornell and Wieman received the 2001 Nobel Prize in Physics for this work, along with Wolfgang Ketterle of MIT who had also created a BEC.

Only a few years after the creation of that first BEC, in 1999, Lene Hau, a Danish-born physicist working at Harvard and the Rowland Institute in Cambridge, Massachusetts, published a breakthrough paper in the prestigious research journal *Nature*. Hau had become interested in BECs and used sodium atoms to make some of her own in 1997. She planned to explore them using laser light, but then pursued the properties of the light itself in the condensate.

The result was spectacular. Despite its matter-of-fact title "Light speed reduction to 17 metres (*sic*) per second in an ultra cold atomic gas," her research made the front page of the *New York Times* and news outlets worldwide, for Hau's team had reduced the speed of light by a factor of 18 million. For the first time ever, light propagated at a speed comprehensible to humans; 17 m/sec is 61 km/hr or 38 mph, attainable by a fit bike rider and comparable to the fictional value 10 m/sec in *Redshift Rendezvous*. Hau's group did this by using a technique called electromagnetically induced transparency (EIT) to alter the optical properties of a cold atomic gas. (Hau was not the first to so use EIT. Stephen Harris of Stanford, a coauthor of Hau's paper, had used EIT in 1995 to slow light, but by a factor of 100, not millions).

EIT makes a dense atomic gas temporarily transparent. The gas is ordinarily opaque because light is absorbed as its electric field strongly couples to the electrons in the atoms and makes them oscillate. That takes energy from the light, and it dies away before it travels very far through the gas. But by using light from two laser beams set exactly out of phase so they cancel each other, the net motion of the electrons can be set to zero and the gas becomes transparent. To make this work, the laser frequencies and the quantum energy levels of the atoms in the gas must be carefully matched to get the cancellation effect, which in quantum physics is called interference. That's after making the gas ultracold, itself not easy to do.

Hau has described the whole experiment as being "just on the borderline of what was possible." In a marathon 27-hour session, she first cooled a collection of sodium atoms with precisely known energy levels to billionths of kelvins. Then she applied a "coupling" laser beam at a wavelength of 589 nm, and sent a second "probe" beam at almost the same wavelength through the now transparent gas, where it propagated at a dramatically low speed.

Hau determined the speed using a nano-sized version of the experiment Galileo had proposed centuries before, measuring time of flight over a known distance — in this case, the minute spatial extent of the atomic gas. For one measurement cited in the paper, the light covered 229 μm in 7.05 μsec, giving a speed of 32.5 m/sec. However, Hau obtained her lowest speed, 17 m/sec, when she made the gas denser by cooling it below 435 nanokelvin to turn it into a BEC containing one to two million atoms.

To understand why the speed was so low, it's essential to recognize that Hau measured a group velocity, not a phase velocity, because her laser produced pulses a few microseconds long, not a continuous beam of light. As established by the French mathematician Joseph Fourier in 1823, any pulse can be expressed as a sum of infinite sine waves, each with its own frequency, and carefully chosen to interfere with each other to produce the particular pulse duration and shape. In essence, a pulse is a packet of waves of different frequencies, and so has a group velocity when it travels through a medium.

If the medium displays strong dispersion — that is, if the phase velocity changes sharply with frequency — the group velocity can be very different from c. The group velocity is determined by an effective group refractive index, which is the usual refractive index with an added contribution due to dispersion that can be positive or negative. In Hau's experiment, the strong dispersion at the EIT condition produced a large positive addition that brought the group refractive index to the tens of millions. Since the group velocity in the medium is c divided by the group index of refraction, that resulted in very small group velocities.

Like Bob Shaw's fictional slow glass, Hau's BEC reduced the speed of light by a huge factor, though Hau's slowest speed was still orders of magnitude faster than Shaw's. If Hau's sodium BEC could be made

6 mm thick like a piece of slow glass, at a light speed of 17 m/sec it would hold only 350 μsec of beauty or memory, not the ten years in Shaw's story.

However, in 2001, Hau went slow glass one better; she managed to stop light completely in its tracks, then later recover it and send it on its way. When she turned off the coupling laser in her experimental setup, the BEC became opaque; the light pulse was no longer being transmitted and it disappeared. But when she turned the coupling laser back on, the BEC released a light pulse identical, in phase and amplitude, to the original pulse. The quantum details of the laser light had been stored or imprinted on the quantum state of the BEC. The storage time was only a millisecond, but in 2009, Hau improved the time a thousandfold, holding light for over a second — another way to delay light.

Also, in 2007, Hau and her colleagues showed that information could be passed from light to matter and back to light; they stored a light pulse in a BEC, then allowed those atoms to move to a second BEC. When the coupling laser was turned on, the second BEC released a light wave identical to the original one stored in the first BEC because the migrating atoms carried the necessary quantum information.

Hau's work has helped open up a whole new field in which researchers explore what light can do in exotic media. One important outcome is the discovery that slow light can be produced in media other than BECs, which avoids the difficulties of creating and maintaining ultracold atomic gases. Slow light has been generated in hot atomic vapors, which are easier to produce than extremely cold ones, and also in certain dispersive materials at room temperature. One of these is a particular type of glass optical fiber (called "erbium-doped") already widely used in telecommunications, which is especially encouraging for practical use.

Researchers have also found that besides slowing down, light in a medium can display startling properties that seemingly violate relativity and causality. In theory, a pulse can have a group velocity that is greater than *c* or negative, in which case it's called "fast light." Negative group velocity describes a pulse that travels opposite to the direction of entry into the medium. This backwards light is extremely counterintuitive. For a typical hump-shaped pulse that smoothly falls away from a peak to its leading and trailing edges, theory predicts that the peak can leave the far end of the medium *before* entering the near end and sooner than if it had traveled the same distance in vacuum, while sending another peak backwards.

Like slow light, these strange behaviors arise from the group index of refraction, the ratio of c to group velocity in the medium. A value greater than one, as found for an ordinary medium like glass, gives a speed less than c. By the same definition, a refractive index between zero and one corresponds to a group velocity greater than c, up to infinity; and a negative index represents a negative group velocity or backwards light.

The theoretical finding that group velocity could apparently violate relativity by exceeding c greatly concerned physicists soon after Einstein's theory was announced. Much later, the advent of lasers and of methods to produce extreme values of the group refractive index — the same approach based on strong dispersion that led to slow light — allowed researchers to explore the theory, not always in atomic gasses. In 1982, S. Chu and S. Wong at Bell Laboratories confirmed the theoretical predictions by sending light pulses through a piece of semiconducting material. They set up refractive indices that gave the pulses group velocities greater than c, negative, and even infinite — in which the peak of the pulse leaves the far end of the medium just as the incoming peak enters the near end, thus seemingly traveling the length of the medium in zero time.

More recent results extend these observations. In 2000, scientists at the NEC Research Institute obtained a refractive index of –310 in caesium vapor and saw superluminal behavior. In 2003, Michael Stenner and Daniel Gauthier at Duke University and Mark Neifeld at the University of Arizona sent a light pulse through potassium vapor in less time than it would take to cover the same distance in vacuum, indicating a group velocity greater than c. In 2006, Robert Boyd and his team at the University of Rochester produced a refractive index of –4,000 in a long piece of optical fiber. This corresponded to a backwards pulse with the low group velocity of 75 km/sec, which — for the first time ever — the researchers directly observed as it traveled along the fiber. The researchers also saw the peak of a pulse emerge from the far end of the fiber before the incoming peak entered the near end.

Such results raise eyebrows because unlike phase velocity, group velocity has real physical meaning. But group velocity is not signal velocity, the speed of information transfer. Since it takes change to convey information, the part of a pulse that varies smoothly and predictably does not carry information. Rather information is conveyed by sharp changes, such as at the pulse's leading edge where it first turns on. Even if the pulse's peak moves faster than c, the leading edge does not, because the electrons in the medium have inertia and cannot respond to the leading edge the very instant it arrives. Hence the light is initially unaffected by the medium and moves no faster than c, preserving the relativistic speed limit for information transfer. Experimental support for this view came when Stenner, Gauthier and Neifeld sent pulses representing "0" and "1" bits through their potassium vapor. They found that this information traveled more slowly than c even when the peak of the pulse was superluminal.

There's a related explanation for the discomfiting result that for a negative value of n, the peak emerges from the medium just as or even before the incoming pulse enters it, which seems to violate causality. But the paradox is only apparent, because the pulse is extended in time. The leading edge of the entering pulse reaches the far end of the medium before the peak enters it, and provides information that allows the entire pulse with its peak to start rebuilding at the far end. This produces both the emerging and the backward pulses.

Nevertheless, fast light and backwards light seem almost eerie, and not all scientists are convinced that we fully understand these remarkable effects and their relation to the transfer of information. But at least the phenomena don't come from quantum weirdness. They arise from well-established classical physics that treats the wave nature of light and its interaction with matter. What is relatively new is techniques to produce highly dispersive situations, which produce atypical values of *n* and seemingly odd behavior (in Chapter 6 we'll encounter other ways to generate exotic values of *n*).

Besides the scientific interest in observing and explaining these striking effects, it's natural to ask what impact they'll have on technology and daily life. Uses for fast and backward light aren't yet obvious, but it's not too early to project what slow and stopped light can — or cannot — contribute. One application we won't be enjoying anytime soon is the storage of beautiful scenery in a pane something like Bob Shaw's slow glass, for the time delays so far are only seconds, not days or years. Also the experiments to date do not operate on a whole image, but only on a pulsed ray of light, with one exception. In 2007, John Howell and his group at the University of Rochester successfully slowed an image of their institution's initials "UR," delaying its transmission by up to 10 nsec. The figure was only a fraction of a centimeter across, but this experiment suggests future possibilities for slowing or stopping entire images.

The real promise of slow light, however, is not that it may someday replicate the science fiction idea of slow glass. It is that it will add new capabilities to the technology of light or "photonics," whose use is growing rapidly. It shows up in making and playing CDs (compact discs) and DVDs (digital video discs), in television and computer displays, and in high efficiency energy-saving light sources. Light also contributes a great deal to data handling though its major role in telecommunications. The world's Internet, message, and data traffic is carried around the globe by pulses of light traveling through optical fibers. Compared to electrons in wires, photons in fibers have higher speed and greater capacity to carry information (that is, they provide higher bandwidth).

But as a newer technology, photonics has fewer ways to manipulate photons than electronic technology has developed for electrons. The ability to slow and stop light would add versatility. For example, suppose two different optical signals in the form of packets of bits are due to arrive simultaneously at a switch in the optical fiber network. Rather than avoid conflict by dropping one packet while the other proceeds, which would be inefficient, a slow light device could delay one signal just long enough to let the other clear the switch first. Slow light methods are already good enough to delay light over time ranges useful for telecommunications.

Stopped light also offers a way to store and retrieve data in the form of light pulses, as we now do for data in electronic form. When Lene Hau reported her stopped light experiment in 2001, this application was noted in the very title of the paper, "Observation of coherent optical information storage in an atomic medium using halted light pulses." Turning off the coupling laser to store the information in the pulse within the atomic gas, and then turning it on to release the pulse, amount to the "write" and "read" functions necessary for a usable memory. Hau also demonstrated multiple read-outs after a "write," another necessary feature for practical data storage. Nearly 1,000 other papers have cited that work in the decade since it was done, indicating the strong interest in this application.

Data storage is an essential step toward the goal of computation by light, which would use pulses of light rather than electrons moving through chips and wires. Beyond the advantages of eliminating physical connections, such a computer could offer the conceptual breakthrough of quantum computing. This is the direct use of fundamental quantum properties to transfer and manipulate data, in theory far more powerfully than existing methods, and also providing unbreakable security for the transmission of encrypted messages. In quantum computing, "qubits" or quantum bits replace the binary digits or bits now used. Just as bits are embodied in transistors, basically "on" and "off" switches that represent binary "1" and "0," qubits can be embodied in the quantum states of photons (as well as electrons and atoms).

Intriguingly, quantum computing applies the weird quantum properties of superposition and entanglement. Qubits employ superposition, the fact that a quantum object like a photon represents many different values of a physical parameter at once. This comes from the quantum mechanical wavefunction, which is the mathematical description of all the physical properties that a particle can have. However, until a property is actually measured, it's as if all the possible values of that property are in a kind of suspension; they are only potentially real until the act of measurement pulls out one definite value from the sheaf of possibilities.

In the next chapter, I'll say more about the strange property of superposition, but a concrete example shows how it works to allow photons to represent qubits. The electric field associated with light can be polarized, which means oriented either vertically or horizontally. Under the right conditions, a single photon has an equal probability of being in either condition, which can be interpreted as representing binary "1" and binary "0" simultaneously. Hence a photon can operate as a qubit that carries more information than an ordinary bit — *both* binary 1 and binary 0, rather than *either* binary 1 or binary 0. In ordinary computing, two bits can hold only one of the four numbers 00, 01, 10, 11 (decimal 0, 1, 2, 3); in quantum computing, two qubits can hold all four numbers at once. This means that a quantum computer can deal with many pieces of data simultaneously, a tremendous boost to computational power. According to Harvard chemist Alán Aspuru-Guzik, a researcher in quantum computing, a machine working with just 150 qubits would equal the combined computing power of all the world's supercomputers.

Entanglement is the other unique feature of quantum computing. It is the mysterious property by which two quantum objects like a pair of photons or electrons, once linked, remain somehow connected even if they are then physically separated, no matter how far apart. If the properties of one change, the other responds faster than can be accounted for by a speed-of-light transmission between the two. This is the phenomenon Einstein called "spooky action at a distance," which contributed to his mistrust of quantum physics. Entanglement allows information to be securely transmitted from one location to another by "quantum teleportation," a method to transmit the properties of a quantum object so it is exactly reproduced elsewhere.

Features like entanglement and teleportation seem more science fiction than science, as if we now know how to beam objects and people through space in the best *Star Trek* tradition. But they are real, well-documented science, although far from being fully understood. Nevertheless, as I'll discuss further in the next chapter, scientists and engineers successfully apply these mysterious quantum principles to manipulate data carried by photons in the real world.

The technology of slow and stopped light is not *per se* part of quantum computing, but combines with it to move quantum computation a step closer to practicality. In 2008, researchers at the California Institute of Technology stored light in an atomic gas and retrieved it much as Hau had done. The novel features were that the light consisted of two entangled photon states which created entangled atomic states, and that the entanglement survived both the "write" and "read" operations. This would make it possible to use both photons and atoms where each serves best in a quantum information network.

Though slow, stopped, fast, and backwards light have their strange aspects, careful analysis shows that they are after all understandable. With that in mind, their weirdness quotient doesn't begin to compare to light's quantum strangeness. Much of what we do know about superposition and entanglement comes from studies of single photons, which are extreme states of light — extreme in demonstrating its particulate nature, and extreme in intensity as the smallest possible unit of light. At the opposite pole are natural and human-made sources that pour out photons in unimaginable numbers, from our own Sun to lasers that deliver stunning intensities. Both extremes deserve their own discussion. In the next chapter, I'll compare floods of photons to single ones, and consider one, two or three photons at a time to say more about entanglement, superposition, teleportation, and quantum computing.

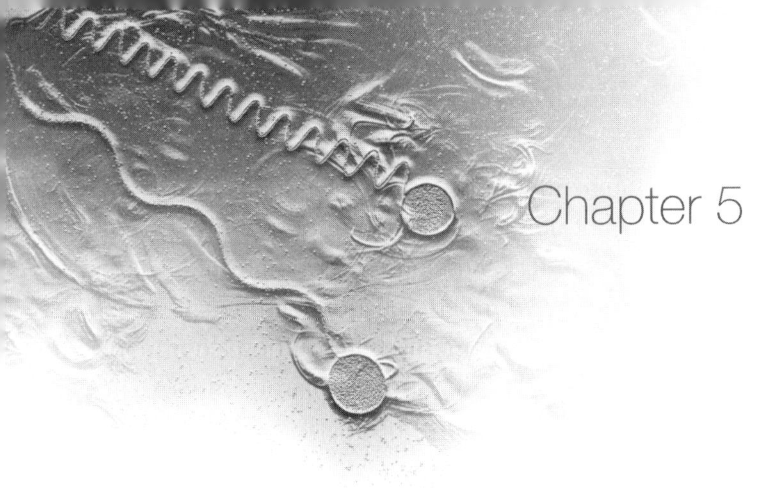

Extreme and Entangled Light

The universe flashed into existence with an inconceivable outpouring of energy that marked the Big Bang. What started it all is still an open question that may never be answered, but there is considerable evidence about the state of the universe a tiny instant after the (literal) beginning of time. What was then a small, extremely hot cosmos contained photons with incredibly high energies. As the universe expanded, the temperature decreased and the nature and distribution of the light changed. Some turned into matter that eventually became stars and other light sources that now stud the universe; some continued to travel through space.

Our own Sun is one of those stellar sources, producing over 10^{26} watts of radiation or around 10^{45} photons/sec; enormous, but puny compared to much more powerful supernovas and gamma ray bursts in the distant cosmos. On the other hand, the intensity of the light that filled space soon after the Big Bang diminished as the universe grew. That light has now become the cosmic background radiation at a low density of 400 photons/cm^3. Also the vast distances in the universe can make even blindingly intense sources like supernovas look dim. A star at the farthest limit of naked eye observation delivers only a few hundred photons per second into the eyes, an infinitesimal flux. And whenever an electron in a single hydrogen atom, isolated in interstellar space, drops from a higher to a lower energy level, it emits just a single photon.

Nothing human-made could come remotely close to conditions right at the Big Bang, but we can recreate what they were like soon after that instant. At the opposite extreme, we have also learned to make and use light one photon at a time. The range in intensity that technology can achieve is remarkable; even an ordinary 100 watt light bulb pumps out over 2×10^{20} photons/sec, whereas a source producing one photon per second generates only about a nanonanowatt (this low level explains why we don't experience light as composed of individual photons). What's also surprising is that for all the impressiveness of high intensity light, the consideration of one or two photons at a time has truly profound scientific implications — and also delights science fiction fans, for it leads to what used to be pure fantasy, teleportation and quantum technology.

A single photon is hard to detect, but human-made light at huge scales, the Big Science of light, is easy to spot; for instance, at the National Ignition Facility (NIF) of the Lawrence Livermore Laboratory in California. If you were to visit the installation, which covers the area of three football fields, you'd see thousands of huge optical components such as lenses more than a meter across, backed by rack upon rack of electronics and power sources as well as multiple control stations. These suggest that something extraordinary is going on with light, and indeed it is. You'd be standing in the middle of what is planned to be the world's most energetic laser installation, completed in 2009 at a cost of $3.5 billion. At full blast, its 192 merged laser beams are designed to produce a peak power of 500 terawatts (5×10^{14} watts) at an ultraviolet wavelength of 351nm, in brief pulses nanoseconds long.

This is one extreme of light, a source powerful enough to replicate conditions in the early universe and inside stars with temperatures of millions of degrees and pressures of billions of atmospheres. NIF's aims are as impressive as its size and power. One goal is to ignite nuclear fusion in hydrogen to produce helium and release a flood of energy, more energy than it took to initiate fusion. This would be like bringing down a piece of the Sun to generate limitless energy, as hydrogen fusion in the Sun has done for five billion years. After sixty years of unsuccessful tries at practical fusion power, it's hoped that the method at NIF, called inertial confinement fusion, will be a big step toward that goal (other different approaches also continue, such as the European ITER project, now under construction; it will use magnetic fields to confine the hydrogen as it is heated to the point of fusion). Other goals at NIF are to study supernovas and other astrophysical process and to examine what happens inside nuclear explosions.

Though the grand scale of the NIF is overwhelming, a tiny single-photon light source has its own kind of impressiveness. Instead of great power, it holds a great mystery — the mystery of what photons really are and what their quantum properties mean. It's remarkable that the technology of light can span this extreme range in power, primarily through the use of lasers but also through other types of sources that are part of a new photonic technology. Although lasers were invented 50 years ago, far from becoming dated, they constantly become more embedded in daily life and in applications in commerce, art and entertainment, warfare, and science. With their enormous versatility, it's no surprise that along with other sources, they can span the range from the least to the greatest flow of photons.

The basic principle of lasing uses the fact that atoms occupy separate, discrete quantum mechanical energy levels like rungs on a ladder. An electron in an atom can be excited to a higher rung by an external energy source such as an electrical spark. As the electron returns to its lower level, it releases a photon at a specific wavelength defined by the difference between the two energy levels. This single weak pulse can be amplified and the process turned into a sustained light source if the production of photons is spread from one atom to many and if the supply of electrons on the upper rung is constantly replenished.

In a working laser, sufficient numbers of electrons are held in upper levels by constantly pumping in energy and by judicious choice of atomic energy levels. A chain reaction of photons is produced by using yet one more way in which Einstein enlarged our understanding of light, the idea of stimulated emission that he stated in 1917; if a photon flies by an excited atom, the photon induces the atom to release a second identical photon. In this way, one photon in a collection of atoms can lead to two photons that then make four, and so on. If the atoms occupy an enclosure or "cavity" with a mirror at each end, the photons bounce back and forth, recruiting more atoms with each pass. With a small hole in one mirror, the resulting flood of light exits as a coherent beam, with all the photons at the same frequency and in exact phase. This is a LASER, with its light amplified by stimulated emission of radiation, which is the origin of the acronym.

The first laser, invented in 1960 by Theodore Maiman at the Hughes Research Laboratories in California, used a cavity made of a solid rod of ruby. Solid state lasers are still widely used and the NIF is of this type, using glass slabs that contain the element neodymium as lasing elements to amplify the light. Lasing can also be sustained in a gas such as carbon dioxide, CO_2, which uses molecular rather than atomic energy levels, and even in a liquid. With this range of media, lasers can operate at wavelengths from a few hundred nanometers in the ultraviolet through the visible and infrared to the edge of the microwave region.

Before lasers came along, conventional sources produced bright light at sites such as the Eiffel Tower and the Empire State Building. One of the brightest contemporary sites is at the Luxor Las Vegas Hotel, where 45 Xenon lamps and over 300,000 watts of electrical power generate a light beam aimed straight up that is said to be visible from space. But lasers outstrip such sources in power and intensity. Power is energy per second, and a laser can deliver very high power if its output is compressed into a brief pulse, as with the NIF. Intensity is power per square meter, and there too a laser outstrips conventional sources because its coherent light can be focused to a small region the size of the wavelength. In my own laboratory, a CO_2 laser producing 20 watts, a power level that would result in an extremely dim incandescent lamp, could bore a hole through a thin slab of lead with its finely focused beam.

That destructive ability has made lasers favorite science fiction weapons, going back to H. G. Wells' *The War of the Worlds* from

1898 — not that lasers existed then, but in the story, the Martian invaders of Earth use a devastating invisible heat ray much like the beam of infrared light from a CO_2 laser (in the film version from 1953, the rays were made visible for dramatic impact). The ray guns used by Flash Gordon and Buck Rogers in 1930s comic strips and movie serials also qualify, and the original *Star Wars* (1977, retitled *Star Wars: Episode IV, a New Hope*) features a huge Death Star laser the evil Empire uses to destroy an entire planet.

The science fictional flavor of laser weaponry was so prevalent that in 1983, when U.S. President Ronald Reagan proposed shooting down enemy missiles with lasers, the project's official title "Strategic Defense Initiative" quickly devolved to "Star Wars." Sometimes the science fiction and the reality bled into each other. The Strategic Defense Initiative was satirized in the near-future science fiction film *RoboCop* (1987), and the plot of the film *Real Genius* (1985) features a laser mounted in an aircraft as an assassination weapon — not so different from the present use by the United States of armed drone aircraft against selected enemies. Though these do not use laser weapons, they do use laser target designators and laser-guided missiles.

A few fictional uses of powerful lasers are positive. In *Chain Reaction* (1996), though it's never explained why, a laser is somehow essential to extract energy from hydrogen to provide the world with clean power. In *Spider-Man 2* (2004), a physicist uses lasers to ignite a fusion reaction, also for the good of humanity. Despite these plots that foresee what the National Ignition Facility hopes to do, neither film resists the temptation to display spectacular explosions from the fusion processes — fortunately, affecting the villains rather than the heroes — before settling down to tout the benefits of laser induced fusion.

Science fiction filmmakers will probably never find it dramatic to show the opposite of powerful, destructive lasers, namely, tiny sources producing only one or two photons; yet these may prove as important as the huge sources to extend our understanding of quantum physics and underpin quantum technology. A variety of methods exists to produce the minimalist output of a single photon, though none is perfect. One is to attenuate the output of a laser until the average output per pulse is about one photon. But because of statistical variation, some pulses will contain no photons, which is a loss of efficiency, or more than one photon. Another approach is to isolate a single atom and use single photons from its electronic transitions, but this is technically difficult. Less demanding is to use the emission from atoms held as impurities in solid materials, such as silicon and nitrogen atoms embedded in the pure carbon of diamond.

But the most promising type of single photon source for general use is the quantum dot, a kind of artificial atom. This is a semiconductor structure with nanometer-size dimensions, which give its electrons energy levels like those of an isolated atom that can generate one photon at a time. These dots can be made by various processes and are rapidly being developed to serve as reliable sources for single photons.

Over a century ago, even before such devices existed, the quantum mysteries of light were being explored photon by photon. This brought light's two faces into direct opposition: a single photon, unquestionable evidence for light as particle, encountering the phenomenon of interference, unquestionable evidence for light as wave. Early 20th century technology could not produce quantum dots, but in 1909, Geoffrey Taylor, a student of the Nobel Laureate physicist J. J. Thomson at Cambridge, used attenuation to approximate individual photons. He filtered the light from a gas flame until it was as weak as a candle seen over a mile away; so "feeble," as he put it, that on average just one photon at a time entered his experimental enclosure.

The enclosure contained a thin needle that would show a diffraction pattern in ordinary light, backed up by a strip of photographic film to record what happened. Diffraction is another illustration of wave behavior like Thomas Young's double slit experiment, in which light waves bending around the edge of an object interfere with each other to form light and dark regions. After setting this up, Taylor went off on a sailing holiday and left the experiment alone, letting the result slowly build up over a period of 2,000 hours. On his return, he examined the film and found a perfect wave-like interference pattern that somehow arose photon by photon.

Such results, easy to describe but difficult to explain, unmistakably present the quantum enigma that physicists still struggle to understand. Richard Feynman, whose extraordinary insight into the quantum behavior of photons earned him a Nobel Prize, has called the behavior in the double slit experiment "impossible, *absolutely* impossible, to explain in any classical way, and which has in it the heart of quantum mechanics. In reality, it contains the only mystery" — the mystery of duality. A photon must go through one slit or the other like a particle, yet the interference pattern shows that it goes through both slits like a wave, or somehow interferes with itself, unlikely as these seem.

Part of the strangeness is that a photon acts like wave or particle depending on the experimental arrangement. Use a screen on which the light impinges, as Thomas Young did, or a strip of film, as Geoffrey Taylor did, and the expected interference pattern of light and dark areas indeed emerges. But monitor each slit instead with a detector that clicks whenever a particle comes through, and sure enough, particle-like behavior results. With multiple runs of individual photons, each detector clicks exactly 50% of the time, and whenever one clicks the other does not, showing that a given photon goes through only one slit.

This puzzling behavior has been observed with purer single photons than Taylor could provide. In 1986, P. Grangier, G. Roger and A. Aspect at Institut d'Optique Théorique et Appliquée, Orsay, and Collège de France, Paris, obtained single photons from a certain atomic transition and sent them through a beam splitter. This is a thin plate that divides a light beam by reflecting part and transmitting the rest, hence giving a photon a choice of ending up on one side of the plate or the other. Detectors placed on both sides exhibited anticorrelation; when one showed a photon, the other did not. Yet when single photons from the same source went through an interferometer, which splits a light beam into two beams that later recombine and interfere, perfect interference patterns were formed as individual photons accreted, just as Taylor had observed in 1909.

The quantum mechanical reality that the observational method determines whether the wave or the particle aspect appears is subject to much interpretation, debate, and philosophical discourse. But if this isn't sufficiently weird, add one more fillip: in the arrangement with two detectors placed on either side of a beamsplitter, put one at a location many kilometers away. That doesn't change a thing. If the nearby detector clicks, the distant one does not; if the nearby one does not, the distant one somehow knows to click, apparently instantaneously, and every time.

This is an example of entanglement, the peculiar, ill-understood phenomenon where two quantum states — in this case, representing two choices for a single photon — are correlated so that a measurement of one immediately determines the other, even if it is far distant. Entanglement can also happen between two photons and also for other quantum particles such as electrons, so that determining the state of one instantly sets the condition of the other. This has deep meaning for quantum physics and also plays an essential role in the quantum technology now being developed, but Einstein found it too eerie to accept.

Einstein's assessment of entanglement goes back to a classic, highly abstract paper from 1935 that he wrote with two young colleagues, B. Podolsky and N. Rosen, which is so famous that it is known simply as EPR. In an era when quantum physics was new and its implications were widely discussed, EPR's title posed a fascinating question. "Can Quantum-Mechanical Description of Physical Reality be Considered Complete?" the authors asked, and came up with a resounding "No!" based on what they saw as a contradiction.

EPR postulates a *Gedanken* or "thought" experiment, Einstein's favorite way to explore nature, by considering two quantum systems that start out as interacting or connected, but are then separated. This could be realized, for instance, with two electrons arranged to have the same speed in opposite directions so that they steadily move apart. As intuition tells you and as the paper shows in detail, given how the particles start out, if you measure the location of particle 1 to some particular accuracy, you immediately know the location of particle 2 to the same accuracy. The same is true for momentum; measure it for one particle and you immediately know its value for the other.

This is troublesome for a couple of reasons. One is that it upsets the Heisenberg Uncertainty Principle. As I wrote earlier, this cornerstone of quantum physics says that certain pairs of physical quantities, such as position and momentum, can't be simultaneously measured to infinite accuracy because measuring one property (say momentum) changes the other (position) — yet EPR shows that one can indirectly measure the properties of particle 2 without affecting that particle.

Another deep issue and a main subject of the EPR paper relates to the so-called Copenhagen interpretation of quantum physics put forth in the 1920s, mostly by the great Danish physicist Niels Bohr. Unlike classical physics where we calculate definite values for the position, momentum, and other properties of a particle, in quantum physics all we know are probabilities for these quantities. The probabilities are expressed in the mathematical construction called the wavefunction of the particle, which amounts to a table of probabilities. It might say that a particular electron has a 27% chance of being located here, a 16% chance of being there, and so on; a 53% chance of having this value of momentum, a 17% chance of having that value, and so on; and similarly for every other one of its physical properties.

Therefore the wavefunction contains all possible values of a particle's parameters, but none really exist until an actual measurement is made. This is the origin of the notion of superposition, where the particle somehow potentially occupies all its possible states at once, until a measurement selects out one specific state. In the language of the Copenhagen interpretation, that makes the wavefunction "collapse," pulling the measured value out of the multitude of possibilities and rendering all other values irrelevant for that measurement. It's a bit like picking a card out of a deck. Each of the 52 cards can potentially be selected, and with a known probability (for instance, the odds of pulling out a spade are 1 in 4, and the odds of pulling out the three of hearts are 1 in 52), but in the concrete event, only one card truly is. Once that card is chosen, this real, definite result leaves all the other possibilities unrealized until the deck is reshuffled for a new choice.

This Copenhagen interpretation has been remarkably successful in predicting the real-world behavior of quantum systems. But the EPR paper flies in its face, because as the paper argues, if we can know the physical parameters of particle 2 without directly measuring them, then these are real objective properties of the particle, not just a set of probabilities that collapses into one value under the act of measurement. As EPR put it, "We are thus forced to conclude that the quantum-mechanical description of physical reality given by wave functions is not complete."

One response to the reasoning put forth by EPR is the idea of "hidden variables," additional factors that represent an objective reality underlying quantum mechanics, and that when added to the theory would make the wavefunction a complete description of reality. But though physicists and philosophers have yet to reach final conclusions about the relation between quantum physics and reality, the plain pragmatic fact that the Copenhagen interpretation works so well gives good reason to preserve it. Fortunately, it can be squared with the EPR argument, but at the cost of choosing a "non-local" explanation. In this view, each particle retains the full range of possible values in the wavefunction, until carrying out a measurement on one particle of a linked pair instantly collapses the wavefunction for the other particle too, no matter how far away; so going back to the experiment with one photon and two slits, the distant detector for slit 2 knows not to click the instant the detector at slit 1 clicks.

This idea of remote influence is hard to swallow; Einstein called it, you can be sure not in a flattering way, "spukhafte Fernwirkung" or "spooky action at a distance." Nevertheless, it is rigorously testable, as was clarified by the American theoretical physicist David Bohm. In 1951 he gave an especially lucid version of the EPR argument using a pair of electrons with opposite values of the quantum property known as spin. Spin is roughly analogous to the classical picture of an electron rotating around its axis like a tiny top, but it is quantized and can take on only one of two values. These are labelled "up" and "down" because spin creates a magnetic field with north and south poles. The spin direction is measured by sending the electron through a magnet that kicks it up or down depending on which way the electron's north pole points.

The stark choice of only two possible states makes it relatively simple to design and perform an experiment to test these issues. It's feasible to create an entangled electron pair with zero net spin, so when one spin is up, the other must be down. To test entanglement, an experimenter could separate the electrons and then use magnets to see how measuring the spin state of one affects the other. However, it needed one more result to make the experiment fully meaningful. That was provided by the Irish theoretical physicist John Bell in a deservedly celebrated result, Bell's Theorem.

In a subtle piece of reasoning, in 1964 Bell considered measurements on the spin states of two entangled electrons. He found numerical limits on the correlations among the results that would be measured if there were no action at a distance, and showed that quantum mechanics violated these by leading to stronger correlations (this would be true even if the quantum theory included hidden variables). If Bell's result could be experimentally verified, it would be startling. As Bell put it, it would mean that in quantum mechanics

> ... there must be a mechanism whereby the setting of one measuring device can influence the reading of another instrument, however remote. Moreover, the signal involved must propagate instantaneously...

All that remained was to test the prediction, and this is where photons enter the story as a better choice than electrons. Although Bohm and Bell analyzed electronic spin, their ideas carry over to polarized photons. These also come in two states, electric field vertical or horizontal, and are easier to manipulate than electrons. Ordinary light is polarized by sending it through a polarizing filter such as is used in sunglasses, which works like a picket fence. With the pickets set either vertically or horizontally, the filter passes only vertical or horizontal orientations of the electric field. The filter can also determine a pre-existing state of polarization, passing photons with electric field parallel to the pickets and blocking those (by reflecting them) with electric field at right angles to the pickets.

A vertically or horizontally polarized photon is logically and quantum mechanically equivalent to an electron in an up or down spin state. And like electrons with spin, photons can be entangled, for instance, by passing them through certain crystals that split light into two correlated streams, one polarized vertically and one horizontally. (Like any quantum property, photon polarization is subject to superposition. After a photon passes through a polarizing filter set at 45° to the vertical, it has a 50% chance of being in either a vertical or horizontal state. That leaves it in a superposition of the two states, making it a qubit that carries both binary 1 and binary 0).

Even with photons rather than electrons, testing Bell's prediction is a demanding experiment. Several researchers have carried it out, but Aspect, Grangier and Roger — the same group that performed the single photon double slit experiment in 1986 — published what is considered the definitive result in 1982. They created pairs of entangled photons with correlated spins from particular transitions in a gas of calcium atoms, then sent the photons through polarizing filters. Testing all combinations of polarizations for different angles of the polarizers and for many photon pairs, they measured the correlation factor between the polarization states of photon 1 and photon 2. The numerical value they found was 2.7, much larger than the maximum value of 2.0 that Bell had derived for theories without action at a distance and exactly the value predicted from quantum theory.

Other experiments have confirmed this result, and though some physicists remain concerned about possible loopholes in the experimental protocols, the consensus is that the 1982 experiment constituted the death of locality. Quantum mechanics absolutely supports the strange notion that a measurement on one of two linked quantum particles also establishes the state of its partner particle no matter how distant — entanglement is real, and can be demonstrated over substantial distances outside the laboratory. For example, in 2007, Anton Zeilinger of the University of Vienna and his colleagues demonstrated entanglement between photons across 144 km of open space between La Palma and Tenerife, two of the Canary Islands.

This does not mean, though, that entanglement is understood. Is it an intrinsic property built into the very nature of quantum mechanics, or is some influence being transmitted between particles at a certain speed, perhaps infinite? Erwin Schrodinger, one of the founders of quantum mechanics, thought the former. He originated the term "entanglement" when he responded in 1935 to the EPR paper, writing that after two systems have interacted and then separated:

> they can no longer be described... by endowing each of them with a [state] of its own. I would not call that *one* but rather *the* characteristic trait of quantum mechanics, the one that enforces its entire departure from classical lines of thought. By the interaction the two [quantum states] have become entangled.

We still don't know if some kind of transmission is involved, but if it is, in 2008 a group from the University of Geneva set a lower limit on its speed. After generating a linked photon pair, they sent one photon over regular fiber optic telecommunication lines to a detector and analyzer east of Geneva, and the other to a similar installation west of the city. The photons were shown to be entangled over a distance of 18 km. Although the researchers could not determine if the entanglement effects were instantaneous, their measurements showed that any speed of transmission between the two photons was at least 10,000*c*.

Like the dream of FTL travel, the possibility of very quickly or even instantaneously linking two distant locations greatly tempts science fiction writers. Such an interstellar communications device even has a name, the "ansible," coined by the distinguished science fiction author Ursula Le Guin in her novel *Rocannon's World* (1966), and given a back story in her later novel *The Dispossesed* (1974). "What is an ansible?" its inventor is asked, and replies: "A device that will permit communication without any time interval between two points in space" without the "long waiting... that electromagnetic impulses require." Le Guin, and other science fiction writers like Elizabeth Moon and Orson Scott Card who have adopted the ansible name and idea, see fast communication as essential for galaxy-wide government and commerce.

They're right about that, and indeed entangled states can be used to transfer *something,* best described as quantum information; but disappointingly, even entanglement does not break the light speed limit. The process of passing on the information is called quantum teleportation. It was first presented as a theoretical possibility in 1993 by Charles Bennett of IBM and colleagues at the University of Montreal, Technion-Israel Institute of Technology, and Williams College. They deliberately chose the word "teleportation" as "a term from science fiction meaning to make a person or object disappear while an exact replica appears somewhere else." True to this definition, they showed that entanglement can be used to exactly replicate and send the unknown state of a quantum system to a distant receiver, with the original disappearing in the process.

In the science fiction scenario, you'd think one could instantaneously teleport Captain Kirk out of the *Enterprise* by measuring the quantum details of every atom that makes up Kirk, then sending out this detailed blueprint to recreate him at a distant locale; but that can't be done even in principle, because the Uncertainty Principle guarantees that the measurements won't be sufficiently exact. As EPR showed, though, we can avoid the uncertainty limitations by measuring the properties of a quantum object through its entangled partner.

Using this approach, Bennett and his colleagues worked out how to send an accurate quantum blueprint to a distant location. They considered two would-be communicators, Alice and Bob (these names are now enshrined in the teleportation literature), each possessing one of two entangled photons A and B. Using these, Alice can convey quantum information about a third photon X with unknown properties to Bob's photon B with perfect fidelity, so turning B into X. The trick is for Alice to measure X along with A, which entangles the two. Since A is entangled with B, this also correlates B with the combined state of A and X.

There's one more step to complete the teleportation of B into X. With two possible polarization states each for X and A, the analysis shows that B has equal probabilities to end up in any of four polarization states. One is identical to X; the other three are related to it in a way that can be determined, and that's where the speed of light comes in. Bob can learn the additional fact he needs to turn B into X (for instance, change its polarization by 90°) if he knows how Alice's measurement of A and X came out. She can tell him that only by conventional means, whether text message, phone call, or email — none of which can travel faster than *c*.

This theory of teleportation was confirmed in 1997 by Anton Zeilinger, then at the University of Innsbruck, and colleagues, in an experiment that closely followed the theoretical protocol to successfully teleport the polarization state of an X photon. It was technically difficult, using pulsed laser light passing through a crystal to produce the entangled A and B photons, and requiring careful timing to entangle X and A. In 1998, D. Boschi and colleagues at the University of Rome, Cambridge and Oxford Universities, and Hewlett Packard, demonstrated a variant form of quantum teleportation. They teleported the polarization state of one of two entangled photons rather than bringing in a third unknown "X" photon to interact with the entangled pair.

In these experiments, Alice and Bob were separated only by laboratory dimensions, but later efforts have demonstrated teleportation over larger distances. In 2003, researchers at the University of Geneva and Aarhus University, Denmark, teleported photons between two nearby laboratories but through 2 km of ordinary optical fiber. In 2004, the University of Vienna group reported teleportation through 0.8 km of optical fiber laid in the real-world environment of a tunnel under the Danube River. Most recently researchers from the University of Science and Technology of China, and Tsinghua University, Beijing, demonstrated teleportation over a distance of 16 km in open space without using optical fiber, opening up further possibilities for quantum communication.

Teleportation exists, and not just for photons. It can be applied to other quantum systems and several researchers have teleported atoms.

Though the process doesn't defeat the relativistic limit to let engineers design instantaneous ansibles, it still has plenty of interesting applications. As I wrote earlier, the vertical and horizontal polarization states of a photon can represent the binary digits 1 and 0, and a photon that has passed through a polarizer set at 45° to the vertical simultaneously carries 1 and 0. That makes it a qubit, which can be teleported. This ability to transfer information carried on photons opens up the new fields of quantum cryptography, communication, and computing — or at least potentially opens up, since there are many technical obstacles to be overcome for these applications. For instance, quantum properties tend to be lost once the system interacts with the ordinary real world, a process called decoherence.

However, quantum cryptography, the use of quantum systems to guarantee secure communication, is already sufficiently developed for commercial use and also offers initial steps toward quantum computing by light. You might think cryptography is an esoteric process limited to the exotic world of international spying. But with ever more of the world's business carried out via telecommunications networks and the Internet, methods to transmit data in secret are vital to protect transactions, from consumer Internet shopping to large financial transfers.

In conventional cryptography, as is used for Internet credit card transactions, information from the sender is run through a mathematical encryption algorithm that uses a secret key to put it into unrecognizable form. In the encryption standard widely used on the Internet, the key is 128 or 256 bits long. The recipient feeds the same key into a decryption algorithm to decode the information. Considerable ingenuity has been expended to make encryption algorithms that are extremely difficult to break without the key, even with massive computational power; nevertheless, the vulnerability in this system is that the key must somehow be secretly transmitted from sender to receiver. For some transactions this is done by hand, with the key carried by a courier, which has its own vulnerabilities.

The problem could be avoided and an encrypted message sent even over a public channel if the key could be communicated with absolute security. That's what the best known form of quantum cryptography, called quantum key distribution (QKD), does. It is a way to use a quantum connection between two users — Alice and Bob again — to send a random string of bits over any kind of communications channel for use as a key. Thanks to the Heisenberg Uncertainty Principle, any attempt to interfere with the key can be discovered. If Eve (for eavesdropper) attempts to intercept or modify what is sent from Alice to Bob, that amounts to a measurement of the quantum system, which disturbs it: hence the interception can be detected.

There are several variants of QKD, but the original version proposed in 1984 by Charles Bennett of IBM (who later worked on quantum teleportation) and Gilles Brassard of the University of Montreal, known as BB84, gives the basic idea. Alice and Bob each have polarizers that can be set vertical or horizontal, called the VH basis; and at 45° to the left or right of vertical, called the diagonal or D± basis. Alice creates a photon that goes through one of the four polarizer settings at random. She records the measured bit value and basis, and sends the photon to Bob. Bob measures the photon with either basis randomly chosen and records the resulting bit and basis. If Bob's measurement used the VH basis and so did Alice's, then Bob's measurement returns the same bit value; but if Alice's measurement was in the D± basis, then Bob's measurement has a random 50% probability of returning either a 1 or 0.

The process is repeated many times, giving Alice and Bob a list of pairs of bits and bases. Then they compare notes by a regular communications channel, even a public one such as radio or the Internet, and discard results where the bases differ. That leaves half the bits to function as a shared key that is absolutely random, based on the inherently random nature of quantum mechanics (the very randomness, by the way, that was part of what Einstein found unsatisfying about quantum mechanics). Moreover, the Uncertainty Principle guarantees that eavesdropping by a third party changes the photon polarizations, which would show up as errors in Bob's measurements. And even if Eve determines some of the bits, Alice and Bob can combine bits to make a new key that is shorter but with proportionately a smaller fraction of it known to Eve.

Other variants of quantum key distribution use entangled photons, and two quantum channels instead of a quantum and a regular channel. All, however, share the features of producing random keys on the spot and providing evidence of tampering. Quantum cryptography has been demonstrated outside the laboratory over substantial distances, such as a joint effort by Cambridge University and Toshiba which sent a key over 20 km of optical fiber at a rate of 1 megabit/sec and also over 100 km though at a slower rate. Several companies supply commercial quantum cryptography. The method has been used for bank monetary transfers, and in 2007, to transmit voting results for a Swiss national election. Also, in 2008, quantum cryptography was implemented over optical fiber to service an entire computer network spanning several locations within the city of Vienna and extending out to another town.

Entangled photon states, collapsing wavefunctions, single photons with enigmatic quantum behavior — these bizarre properties of light stretch our understanding, yet are beginning to change our world. The day may come when we'll use deep principles of quantum physics every time we talk on our cell phones, an outcome that the pioneers of quantum mechanics probably could not have foreseen. We'll return to these matters in the last chapter, but first it's important to remember that the classical 19th century picture of light as electromagnetic waves also has its novel and fascinating aspects. The whole topic of invisibility, now moving from the pages of the Harry Potter stories into reality, is an application of the wave theory of light, modified as the waves move through new kinds of media called metamaterials, which we pursue in the next chapter.

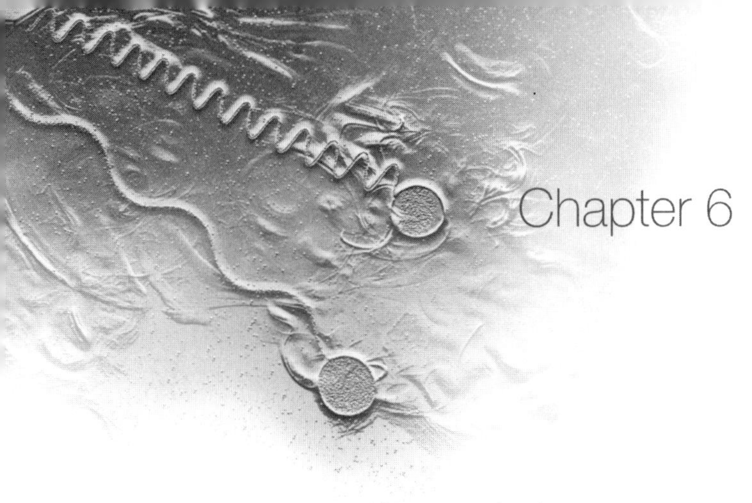

Chapter 6

Invisibility

To be invisible, we imagine, is to be powerful. Envision being able to go where you will without being detected; to secretly gather information, eavesdrop on private conversations, and watch events not meant to be seen; to carry out daring or even immoral acts and escape without a trace. That sense of great, covert, not necessarily legal power explains why personal invisibility and its attractions are such lasting themes.

In real life and in modern times, the uses of invisibility have been less dramatic and more pragmatic. Many applications center around military camouflage, where methods to make a weapon or soldier blend perfectly into the background are a form of invisibility. But the ancient magical power and mystique of the idea still linger, and explain why the steps now being taken toward its real achievement generate vast interest.

For ages, invisibility has played roles in myth, legend, art, and fantasy, often by means of magical devices. Even the Greek gods used this magic when they needed secrecy, as portrayed in the *Iliad,* where the goddess Athena dons a Cap of Invisibility to conceal her actions in the Trojan War. Later, in 380 BCE, Plato recognized the lure of invisibility for mere mortals. In *The Republic,* Glaucon discusses the meaning of morality by telling of the shepherd Gyges in the kingdom of

Lydia, who discovers a golden ring in a cave. Taking the ring to the king's court, Gyges finds that it renders him invisible at will. He seduces the queen, murders the king, and takes over the kingdom. Glaucon uses invisibility as a metaphor to bolster his claim that people behave morally only for fear of penalties for bad acts. If men could become invisible, he says:

> No man would keep his hands off what was not his own when he could safely take what he liked... or go into houses and lie with any one at his pleasure, or kill or release from prison whom he would, and in all respects be like a God among men.

In Shakespeare's 17th century work *The Tempest,* Prospero's servant Ariel commands supernatural powers including invisibility without needing magical props, but these are featured in Richard Wagner's 19th century Ring Cycle. In the first opera of the Cycle, *Das Rheingold,* the chief god Wotan seeks a ring of power forged by the dwarf Alberich from gold stolen from the Rhine maidens. Rhine gold is also fashioned into the Tarnhelm, a magical helmet that makes Alberich invisible when he dons it. This echoes the Nibelungenlied, the 12th century German saga that is one source for Wagner's work, where the hero Siegfried uses an invisibility cloak.

Similar ideas carry over to J. R. R. Tolkien's 20th century classic *The Lord of the Rings,* also constructed around a ring that bestows ultimate power on its wearer including the ability to become invisible. Gollum, the debased character who steals the ring, is similar to Alberich in stature, and in the moral decay induced by the ring and its power of invisibility. Even in today's high-tech world, invisibility plays a role in fantasy. A recent famous example is the Invisibility Cloak in J. K. Rowlings's tales about the young English wizard Harry Potter. The cloak was given to Harry in the first story of the series, *Harry Potter and the Sorcerer's Stone* (1997), and appears in his later adventures.

Beginning in the late 19th century, a different approach to invisibility developed as science replaced magic. Earlier in the century invisibility was still purely magical, for in James Dalton's 1833 novel *The Invisible Gentleman,* a mysterious stranger grants young Bernard

Audrey the gift of disappearance by rubbing an unspecified ointment on his ears. Thereafter Audrey can switch from seen to unseen and back again merely by pulling an earlobe (the magical ointment also sidesteps the vexing problem of what happens to clothes when the body becomes invisible, since it turns them invisible too. It also somehow avoids the even more vexing problem that an invisible person is necessarily blind; scientifically speaking, vision requires interaction with light, whereas invisibility implies no interaction).

But by 1881, in the short story "The Crystal Man," the American writer Edward Page Mitchell was offering invisibility with a scientific basis or at least a scientific vocabulary. Stephen Flack in the story is a young researcher whose scientific mentor subjects him to experiments that alter his skin color "according to well-defined chemico-physiological changes." The next step is to altogether eliminate pigmentation by "absorption, exudation, and the use of the chlorides and other chemical agents." Soon Flack turns translucent, and then he becomes "in the air like a jellyfish in the water. Almost perfectly transparent..." Sadly, he remains trapped in this state, which makes the love of his life heartlessly reject him and leads to his suicide.

Two better known examples that came later may have drawn on Mitchell's work but give more details about the optics of invisibility. In H. G. Wells' *The Invisible Man* (published 1897; film, 1933), a stranger arrives in the English village of Iping, wrapped in overcoat, hat, and dark glasses, his head swathed in bandages, to hide the fact that he is invisible. Eventually Griffin, the mysterious visitor, explains that he is a scientist who has used optical principles to became invisible. "Either a body absorbs light," he says, "or it reflects or refracts it, or does all these things. If it neither reflects nor refracts nor absorbs light, it cannot of itself be visible." As Griffin correctly states, a transparent object with the same refractive index as its surrounding medium, such as a piece of glass (or a jellyfish, as Stephen Flack has it) placed in water, is invisible.

Griffin claims that the materials making up most of the human body are basically transparent, and explains how he has found ways to lower their refractive indices to that of air and to bleach the opaque red pigment of blood. With the help of some mysterious "ethereal vibration" he has used these discoveries to make first a cat and then himself invisible — an achievement, he had thought, that would "transcend magic" and give him power and freedom.

But invisibility is not the glorious state he had hoped. It brings him only ruin and eventually, death — one more example of its corrupting power, especially in the film version, where the Invisible Man is portrayed as an insane megalomaniac who wants to make the world grovel at his feet. Perhaps James Whale, the film's director, remembered that he had directed the classic *Frankenstein* in 1931, because the Invisible Man as scientist is accused of meddling in "things men should leave alone" much like Dr. Frankenstein.

The invisibility in Jack London's "The Shadow and the Flash" (1903) also does no one any good, but the story does offer two different approaches to reaching that state. The tale is cast as a competition between two brilliant young scientists, Lloyd Inwood, tall and dark; and Paul Tichlorne, tall and blond. Life-long rivals in all respects including loving the same woman, they fiercely compete to find the better chemical method to become invisible.

Like Griffin in *The Invisible Man*, Tichlorne opts for perfect transparency. He finds "reagents" that bring about the proper molecular changes to make a dog and then himself transparent, except that something about the condition creates occasional brief rainbow-like pulses of light, the "flash" of the story's title. But apart from this, his method is a success.

Inwood follows another path. A perfectly black object, he says, does not reflect any light and so would be invisible. Seeking "absolute black," he tries a variety of materials, such as lamp-blacks and tars. He finally discovers the perfect liquid pigment, arranges to become coated with it, and truly disappears except for two problems. When he passes in front of an object, the object momentarily disappears, but an observer simply thinks his eyes have blurred. And Inwood can't help casting a shadow. Still the success of his pigment is such that he too displays megalomania and says, "Now I shall conquer the world!"

There is not much world-conquering to be done, however. When the two rivals learn of each other's achievement, they become enraged. Still invisible, they fight viciously in an eerie contest where blows and gasps can be clearly heard, but all that can be seen is a darting shadow and fleeting glimpses of rainbow flashes, until they kill each other.

Given all the ill feeling between and toward invisible people, maybe it's a good thing that more recent fiction is less about personal invisibility and more about concealing objects like spaceships, as in the *Star Trek* television shows and films. In the series, the Romulan Empire, long-time antagonist of the United Federation of Planets which includes Earth, uses "cloaking devices" to hide its predator-like Warbird spaceships. Romulan cloaking first showed up in the episode "Balance of Terror," during *Star Trek's* first television season in 1966. Since the series occurs in a 24th century future filled with advanced technology, there should be a scientific explanation for cloaking, and indeed there is a likely one.

In the series and in various "technical manuals" that entertainingly describe the *Star Trek* universe, we're told that the cloaking device is tied into a spacecraft's deflector shields and needs so much power that it can't operate simultaneously with them. The shields defend against enemy phaser beams by using gravitons, the hypothetical elementary particles that carry gravity. This relates the deflector shields to the distorted spacetime that is connected to a strong gravitational field in general relativity, along with the enormous mass (or equivalently energy) it takes to create the distortion. Putting all this together, it's likely that the deflector shields and the cloaking device use all that power to generate a locally distorted spacetime, which makes phaser beams or light curve around the ship as if it weren't there.

This is just one example of science fiction ideas for invisibility that connect to real science. Lloyd Inwood's recipe for invisibility in "The Shadow and the Flash," complete absorption of light, happens at an extreme level in black holes like the one in the center of our galaxy. Nearer home, stealth military aircraft are invisible to radar partly because they're coated in an "absolute black" that works for microwave wavelengths. Scientists have also used nanotechnology to develop black coatings that resemble Inwood's pigment because they absorb light almost perfectly over a range of visible wavelengths.

A form of perfect transparency as suggested in fiction has also been demonstrated. It's called active or optical camouflage, and it uses a clever display system to paint the front of an object with what lies behind it, giving the illusion that one can see through the object. This could perhaps be integrated into a body suit that provides personal invisibility or at least deflects identification, a popular science fiction idea. In Philip K. Dick's *A Scanner Darkly* (1974), narcotics agent Bob Arctor hides his identity with a "scramble suit." This membrane surrounds him and randomly displays one of over a million different human faces in rapid succession to produce an unidentifiable blur. In *First to Fight* (1997), the opening story in the *Star Fist* series by David Sherman and Dan Cragg, space marines in the 25th century use chameleon-like combat suits that change to match the background. But even these camouflaged warriors might not do so well against the fierce alien in the film *Predator* (1987) who hunts humans and faces off against Arnold Schwarzenegger's military team in the jungle. The alien sports a device that makes him invisible except for a barely perceptible surrounding distortion.

Masking an object by bending light, as in *Star Trek*, has also been realized, but without invoking general relativity. That's good, because it avoids the need for stupendous mass. In the measurement that confirmed general relativity in 1919, as I wrote earlier, the entire mass of our sun (2×10^{30} kg, over 300,000 times the Earth's mass) deflected a light ray from a distant star by only 1.6 seconds of arc or 0.0004 degrees. For an object 100 km away, that would give an apparent displacement from its real position of less than one meter. It's nowhere near enough to curve light completely around a spaceship and hide it — and even this small effect would require that the Sun's mass be placed in the craft. But instead of relativity, cloaking an object by bending light draws on the older theory of Maxwell's electromagnetic equations. Where the novel science enters is in developing new "metamaterials" that control light waves as required.

However, of all the real approaches to invisibility, the most successful so far is the development of stealth aircraft. Its roots and those of other forms of practical invisibility lie in the military art of camouflage, which arose in World War I. The word itself is said to originate from *camoufler*, "to disguise," from French thieves' slang. It was only in the 19th century that various armies took initial steps toward camouflage, when they gave up highly visible uniforms such as British "redcoats" wore for less garish ones. In the mid-1800s British troops in India adopted the earth tone khaki, also used in the Boer Wars of 1899 to 1902.

In that same era, the American artist Abbot Handerson Thayer extended those first steps by studying protective coloration in nature to become "the father of camouflage." Many animals, he noted, display countershading; that is, they are dark on the top half and lighter or white on the underside, which counteracts the shading due to the sun's overhead illumination. Thayer also described two different camouflage styles: blending, where an object or person is made indistinguishable from its background; and disruption, where "the employment of strong arbitrary patterns of color," as he wrote, hinders the eye from following an object's outline.

In 1898, Thayer and an artist colleague suggested adding protective patterns to American warships during the Spanish–American War. Though this was not done, Thayer's ideas along with those of others came into their own in World War I. The realities of aerial observation, trench warfare and U-boat attacks motivated the combatants to make troops, weapons, vehicles, and ships hard to see or at least difficult to target.

Camouflage has been deployed in every conflict since, most obviously, in patterns for uniforms and vehicles designed to blend into woodland, snow, and desert environments ("desert" is the style now seen on U.S. troops in Iraq and Afghanistan). Countershading has been used to make aircraft harder to see from both below and above. "Dazzle" camouflage was applied to warships in both World Wars. These bold high-contrast geometric patterns — often featuring diamond shapes and diagonal lines — did not hide a ship, but followed Thayer's idea of disruptive camouflage to make it difficult to judge the craft's position, course and speed.

A widely circulated story coming out of World War II has it, though, that in Fall 1943, the United States Navy achieved the ultimate in camouflaging a warship. Navy scientists supposedly used Einstein's theories to make the destroyer *Eldridge* invisible (and on top of that, teleported the ship from the Philadelphia Naval Shipyard to Norfolk, Virginia, and back). This legend, called the Philadelphia Experiment (it was celebrated in a 1984 film of that name and in several books) has been debunked on multiple grounds by the Navy and other sources.

However, stealth technology has achieved something similar for warplanes, though designed specifically for how radar works, as indicated by its acronym which stands for "radio detection and ranging." In human vision, the reflected light by which we mostly see originates from diffuse sources such as the Sun or artificial illumination, but a radar source produces a directed beam. This is composed of radio waves in the microwave range with wavelengths of centimeters or less. It is continually swept around in a circle like a lighthouse beam, extending out typically tens to hundreds of kilometers.

When the beam illuminates an object, part is reflected or scattered in various directions and part enters the object. A fraction of the scattered energy returns to the beam's origin where it is detected. The direction of the beam gives the object's bearing, and multiplying the beam's round trip time by c and dividing by two gives the object's distance (this shows the usefulness of light's particular speed. If it were very low, a fast aircraft could change position by the time the beam returned. If it were infinite, the distance couldn't be found). Besides producing a "blip" on a screen that represents the object, some radars such as those seen in TV weather reports measure the Doppler frequency shifts between outgoing and returning beams. This gives the object's speed along the line between it and the radar transmitter.

The amount of energy an object reflects back depends on its size and shape, and on its constitution. For weather disturbances, different amounts and kinds of precipitation reflect differently so the radar return can be interpreted accordingly. For man-made objects, wood, fiberglass and plastic are non-reflective and even transparent so they are "radar invisible," whereas metal is strongly reflective and highly visible. That's desirable for keeping track of commercial airliners, and for sailors. Even a fiberglass sailboat can be seen on radar if its skipper mounts a metal reflector on the masthead.

It's different, however, for warplanes, which ideally should be hard to detect, but can be all too visible to radar. Military aircraft are mostly metal and have protruding parts like vertical stabilizers that reflect microwaves right back to the transmitter. That can produce large radar cross sections, the measure in square meters of how evident the object is on a radar screen. During the Cold War, the United States Air Force B-52 Stratofortress bomber was easily detectible with its huge "side of a barn" cross section of 125 m^2. In the film *Dr. Strangelove* (1964), the classic dark comedy about inadvertent nuclear attack during the Cold War, one of these aircraft penetrates Soviet airspace only by flying very low to avoid enemy radar.

Later designs were less visible; the B-1 bomber from the 1980s had a radar cross-section of only about 1 m^2. But military planners wanted even smaller radar signatures. The key to achieving this was James Clerk Maxwell's 19th century theory with modern twists. Maxwell and others had used his equations to calculate how symmetric shapes like spheres scatter electromagnetic waves. What was lacking was a way to calculate the radiation scattered from the irregular shape of an aircraft, and then to find the shape that would return the smallest possible signal. The answer, ironically in that Cold War era, came from the Soviet scientist Pyotr Ufimtsev, an electromagnetics expert who published a technique to find the scattering from complex shapes.

In 1975 engineers at Lockheed Aircraft's Advanced Development Projects establishment, popularly known as the "Skunk Works," began using this method to produce a stealth fighter aircraft on the computer and then in reality. The result was the U.S. Air Force F-117 Nighthawk, first flown in 1981. Its distinctive multi-faceted appearance resembled a Picasso Cubist painting, sharply different from the sleek, rounded look of most aircraft. Its prototype was called the Hopeless Diamond, because it looked so utterly non-aerodynamic that flight seemed impossible.

The strange shape was due to the limitations of 1970s era computers that could deal only with two dimensional plates to design the fighter, but the shape makes it possible to see how the design works. For instance, instead of a vertical stabilizer, the craft has combined vertical and horizontal stabilizers tilted outward in a V-shape. This directs a reflected radar beam downward and away from the direct line back to its source.

Although the shape gave a tiny radar cross section of about 0.02 m^2, the intuition about the Hopeless Diamond was right; the Nighthawk was aerodynamically unstable. To fly without crashing, it required constant trimming of its control surfaces and therefore its flight characteristics by on-board computers. Later, the stealth B-2 Spirit bomber and F-22 Raptor fighter, first flown in 1989 and 2005 respectively, came from better design computers and displayed rounded rather than faceted surfaces. But they still required control by built-in computers or "fly by wire" for stable flight.

Besides optimizing the scattering profile, these aircraft further reduce their radar cross section with Inwood's method in "The Shadow and the Flash," which is to absorb rather than reflect radiation. The aircraft's metal surfaces are coated with radar absorbing material, for instance, paint containing carbon or metal particles that extract electromagnetic energy from the incoming waves and turn it into heat. That slightly increases the aircraft's surface temperature, with the undesirable side effect of making the craft easier to detect by the infrared sensors that missiles use. Reducing the infrared signature represents a second type of invisibility that needs to be considered for stealth aircraft.

But radar invisibility *per se* has been achieved to a remarkable level. Compared to that enormous side-of-a-barn 125 m^2 cross section of an old B-52 bomber, the combination of shape optimization and radar absorbing coatings has resulted in modern stealth aircraft with various radar cross sections said to be the size of a golf ball, a marble, and a bird or large insect.

The achievement of nearly total radar invisibility is a technological triumph. Its lessons, however, can't be used to create the traditional invisibility of myth and fantasy, where a person or thing vanishes from view whatever its shape and viewed from any angle. But there is another, real-world approach to the type of invisibility the Invisible Man achieved, namely, a kind of virtual transparency introduced in 2003. It arose when Susumu Tachi (then at the University of Tokyo, now at Keio University) and his group considered the problem of occlusion when mixing virtual displays with real environments. If an image of something meant to lie behind a real object is projected from in front of the object, the object obstructs or occludes the image. This destroys the depth cues needed to make the scene look three dimensional and real.

To fix the problem, Tachi's group used the idea of retroreflection where, unlike ordinary reflection, light coming in over a range of angles is reflected back out directly toward the source. This is what makes a cat's eyes and traffic signs shine brightly in car headlights. It can be achieved by embedding tiny glass beads into a thin sheet or mixing them into paint. In the Tachi technique, called optical camouflage, the front of the object is coated with retroreflective material and a camera records what lies behind the object. Then a projector reflects that background image off a partially transmitting mirror onto the object. With the mirror placed correctly, a viewer looking through it sees the object along with the background image reflected back along his line of vision.

The dramatic result, shown in photos and videos from Tachi's group, is that the object looks transparent as if the viewer were gazing right through it. Buildings, trees, people and vehicles in busy street scenes are seemingly clearly visible through the bodies of models wearing a retroreflective jacket and a wizardly cloak. These images received broad news coverage in 2003 that hailed Tachi as the inventor of true invisibility. However the illusion is imperfect, since the projected image loses some brightness and the outline of the retroreflective garment is visible. Also the illusion requires that the camera and projector be set in place with the viewer looking through the half-transmitting mirror.

These limitations could be overcome by making a Harry Potter-like cloak that creates its own illusion of transparency, even as its wearer moves. The back of the cloak would be coated with light sensors and its front with projectors that send the background image outward, but there are big technical challenges: making the sensors and projectors small and numerous enough to give adequate resolution, managing the visual data so the illusion works as seen from any angle, and running all this in real time with a computer (and power supply) built into the cloak. However, nanotechnology may develop to the point where this can be done; there's even promise for electronic and photonic components that can be embedded into a flexible cloak. I'll say more about this possibility later.

Nevertheless, even within its current limitations virtual transparency is potentially useful. For the military, even with a fixed camera and projector the technique could provide superior camouflage, for instance as a portable screen or enclosure to hide vehicles and weapons. A 2007 analysis conducted for the Canadian military points out applications such as active camouflage panels mounted on tanks (and also notes that the development effort it would take to make individual soldiers invisible is probably not justified).

Another potential application is the "transparent cockpit." This allows a vehicle's driver to "see" through its opaque portions, improving safety by showing the surroundings as if the vehicle were made of glass. In the technique, images from external cameras on the vehicle are recast by a computer as if seen from the driver's viewpoint. Then a projector that the driver wears sends the altered images onto retroreflective material covering the opaque areas inside the vehicle. Striking images from Tachi's group show traffic markings and a bicycle in the next lane as if seen directly through the door and dashboard of a test vehicle.

This elaborate arrangement is a far cry from magical invisibility, but its inventors see it as offering similar freedom. "The ultimate goal of the transparent cockpit," they write, "is that the driver should not feel 'I am operating a vehicle,' but 'I am running or flying by myself,'" and they visualize a helicopter pilot flying in a transparent cockpit as if surrounded by nothing but air. This would be a realization of a fantasy aircraft, the Invisible Plane used by the superhero Wonder Woman ever since her comic book debut in 1942, though the invisibility would work from the inside out only.

But how about "real" invisibility, where light is manipulated so as to truly, not virtually, hide an object or person from human eyes? That, it turns out, is possible in principle, starting from the fact that we have known for centuries how to manipulate light with lenses. Now new technology makes it possible to manage light beyond what ordinary lenses can do, even to the point of providing invisibility, and without using enormous masses or energies. In effect, we've learned how to manipulate space and time as seen by electromagnetic waves without having to distort spacetime itself as general relativity requires.

To understand how this is done, consider an ordinary lens, a piece of glass or plastic with curved surfaces. Refraction at the surfaces bends light rays as they pass from air into the lens and then out again, with the degree of bending determined by the curvature and refractive index n of the lens. A ray traveling from air to glass bends more sharply into the lens, and oppositely when it travels from glass to air. According to these rules, when light rays traverse a lens that is thicker at the center than at the rim, they're bent towards a common point when they leave. That's the focal point, and what I've just described is a convex converging lens — the kind that kids use to burn a piece of paper or wood with the sun's rays. A concave lens, thinner at the center than the edges, makes light rays diverge and so it defocuses. Both types have their uses in optical devices.

These devices are designed with the help of Snell's Law of refraction, the mathematical statement that relates the degree of bending to the value of n, mentioned in Chapter 4. It is named for Willebrord Snellius, a Dutch mathematician who stated it in 1621 — though scientists from Ptolemy to Alhazen to Descartes worked on the problem, and the Muslim scientist Ibn Sahl is credited with having first discovered the law in 984 CE. Lenses were used ages before then, according to evidence that they were known to the ancient Egyptians, Assyrians, Greeks and Romans, but Ibn Sahl was the first to design lenses using the law of refraction. Today Snell's Law is firmly embedded in the optical technology of eyeglasses, camera lenses, and so on, where it is applied to materials with n typically between one and two.

That established law, however, took a turn toward practical invisibility when V. G. Veselago, of the Lebedev Physics Institute, Moscow, wrote an apparently fanciful theoretical paper in 1964. He explored Maxwell's equations in a medium whose electrical and magnetic properties combined to give an unheard-of result, a negative value of n. The condition is that the medium's permittivity ε and permeability μ, which define its electrical and magnetic properties respectively and which together determine n, must both be negative.

Light traveling in such a medium is called "left handed," because its group and phase velocities go in opposite directions, whereas in an ordinary "right handed" medium they go in the same direction. (The terms come from a mnemonic, the "right hand rule," that physicists use to track the directions of the fields and the energy flow for an electromagnetic wave). This is like the backwards light in Chapter 4, but the negative index of refraction there came from strong dispersion induced by a special mechanism such as electromagnetically induced transparency.

Veselago's treatment, however, assumed that the innate refractive index of the medium is negative, and he found that this condition offered intriguing properties. Light entering the medium would be refracted opposite to the usual direction, so a convex lens would defocus and a concave lens, focus; instead of incident photons pushing the material because they carry momentum, they would pull it toward their source; the Doppler effect would be reversed, with light becoming more red rather than more blue as source and observer approach each other, and more blue as they recede; and a plane-sided slab with $n = -1$ would focus light like a lens even without curved surfaces, and would not lose any incoming light due to reflection.

Fascinating as these effects are, they did not receive much initial attention. No known material had a built-in negative value of n and Veselago's theory could not be tested or applied. Decades later, no such natural materials have yet been found. But in 2000 and 2001, breakthrough papers by researchers at the University of California, San Diego (UCSD) reported negative refraction in a new kind of structure, an artificial material or "metamaterial" they had fabricated. Though the results were so unusual that they were initially questioned, other experiments confirmed them and Veselago's predictions.

These metamaterials are carefully designed to go beyond nature with tailored electrical and magnetic properties. The metamaterials used at UCSD were composed of hundreds of components arranged in a regular repetitive pattern in space, like atoms in a crystal but much bigger. The components, which were made by standard microfabrication methods developed for the electronics industry, were of two types. The desired electrical properties came from small strips of copper, whose electrons reacted to the electric field in the incoming light to produce negative values of ε. The desired magnetic properties came from small copper rings with gaps, called split ring resonators. These carried currents caused by the incoming light, which created magnetic fields that produced negative values of μ, but only at specific wavelengths of light.

To make an array like this behave like a continuous optical medium rather than isolated components, the component's dimensions and spacing must be less than the wavelength. For that reason the first observation of a negative value of n was with the comparatively large wavelengths of microwaves, not visible light. The metamaterial used by the San Diego group in 2001 demonstrated a refractive index of −2.7 for microwaves of 3 cm wavelength, much larger than the millimeter-sized copper strips and rings and their spacing.

This showed the power of metamaterial science to create exotic optics and it is now an extraordinarily active field. One thrust is to extend the technology to visible light. It's hard to make intricate metamaterials suitable for the minute wavelengths of the visible but there are other approaches. In 2000, J. B. Pendry at Imperial College, London predicted that Veselago's slab-sided negative index lens could give a higher resolution than a conventional lens, which cannot resolve features smaller than roughly the wavelength of the light. This means that the tiniest objects that can be discerned with visible light are a few hundred nanometers across. But in Pendry's analysis, a thin slab of silver tens of nanometers thick would have appropriate negative properties and could act as a "perfect" lens with unlimited resolution, because it transmits more information about the object than a conventional lens.

Later analysis showed that the perfect limit could not be attained, but researchers have used thin metal films to make enhanced "superlenses." In one example, N. Fang and coauthors, at the University of California Berkeley, used a silver film 35 nm thick to achieve a resolution of only 60 nm under illumination at 365 nm, though this lens did not magnify; but in another report by I. Smolyaninov and colleagues at the University of Maryland, a structure using multiple thin gold layers achieved a magnified image along with a seven-fold improvement in resolution. Improved resolutions would hugely affect visible light microscopy for biology, making viruses and proteins directly viewable for the first time. They could also be technologically important. The number of components on a computer chip is limited by the wavelength used in the photolithographic process that lays out the chip, and the number of bytes stored in a DVD is limited by the wavelength of the lasers that write and read them, so computing and storage technology would benefit from higher resolutions.

However, the feature of metamaterials that really captures the imagination of both scientists and the general public is that they offer a way to create true invisibility. If light rays could be made to curve around an object and then rejoin to continue in their original direction — like a stream of water splitting to flow smoothly around an obstacle and then recombining — the light "downstream" from the object would seem unaffected by anything physical. Actually, to make the object invisible as seen from upstream or any direction, all its interaction with light would have to be eliminated, so the ultimate goal is to "cloak" a given region of space to prevent light from entering it, creating a bubble of invisibility.

Could a metamaterial be designed and built to accomplish this? Two theoretical papers published simultaneously in 2006 provided a way to do just that; one by J. B. Pendry, D. Schurig, and D. R. Smith of Duke University, and the other by Ulf Leonhardt of the University of St. Andrews. Using the technique called transformation optics, their approaches were to mathematically push and pull space as if it were elastic, creating a void of a specific shape where light could not enter, as determined by Maxwell's equations. Then the same distortion is applied to permittivity and permeability, which yields the correct refractive properties for a medium that would act to exclude light from the forbidden region.

Both papers gave specific examples of cloaking. Pendry, Schurig and Smith showed how to electromagnetically isolate a spherical volume by surrounding it with a spherical shell that intercepts any approaching light. Then the refractive behavior of the shell is tailored so that it bends incoming rays into itself, guides them around the cloaked spherical volume, and sends them back out to continue their original trajectories — so in principle, the shell and the sphere it encloses leave light waves undisturbed as seen from outside.

This behavior requires a refractive index that varies with position within the cloak, since no fixed value of n could produce the necessary intricate paths for light rays. As n changes to meet the cloaking requirement, it can take on values greater than one, and also exotic values between one and zero as well as negative ones (though values $n < 1$ represent speeds greater than c, they're phase velocities so relativity is not violated).

Turning such a complex mathematical recipe into a real metamaterial isn't easy, but that was accomplished and cloaking was first demonstrated in 2006 by the authors of one of the original theoretical papers — Schurig, Smith, and Pendry with their colleagues. Like early work on negative indices, this was done at microwave wavelengths to make it easier to build the cloak. This intricate structure had ten concentric rings from 5.4 to 11.8 cm in diameter with thousands of millimeter-size copper split ring resonators deposited on their surfaces. The dimensions of the resonators varied on each concentric ring to provide the correct spatial behavior for n.

The cloak was tested with a copper cylinder 5 cm across that strongly scattered microwaves and cast a shadow when it was uncloaked; but placed within the cloak, its scattering and shadow were sharply reduced at a wavelength of 3.5 cm. The researchers did not expect perfect invisibility because they made some approximations to simplify constructing the cloak, but the test object became much harder to detect. To further check that the cloak worked as planned, the experimenters made images that show microwaves entering it, splitting to bypass the central opening with the hidden object, then reforming and emerging from the opposite side of the cloak in the same conformation as when they entered, just as calculated — and utterly unlike the behavior that would be seen with a conventional material.

This experiment definitively established the theory and practice of cloaking with metamaterials and generated vast interest. International news coverage abounded, with (predictably) hyped headlines such as "How to Make an Invisibility Cloak Like Harry Potter's" (Associated Press), "Science Reveals Secrets of Invisibility" (CNN), and "Harry Potter Invisibility Cloak 'within five years'" (London Daily Telegraph). But besides the popular reaction, this first report of cloaking has been cited by other scientists nearly 800 times since its publication; cloaking is a highly active research area.

And well it should be, because comparisons to Harry Potter's cloak are premature, especially for visible light. Between the difficulty of constructing sufficiently fine-grained metamaterials and the high absorption of metals like copper at these frequencies, that's a continuing challenge. Another challenge is making a cloak operate over a range of wavelengths, not only at one. A third is carrying out three dimensional cloaking. The copper cylinder test was only two dimensional because there was no cloaking along the cylinder's axis.

But progress has been rapid. One important step toward cloaking at visible wavelengths was made in 2007 by G. Dolling, M. Wegener, and S. Linden of the University of Karlsruhe and the Institute for Nanotechnology, Karlsuhe, and C. M. Soukoulis of the Ames Laboratory and Iowa State University. Using nanofabrication techniques, they constructed a square repetitive grid of silver with dimensions in the tens of nanometers, much less than the wavelengths of visible light. Though this was not an invisibility cloak, it did display a negative refractive index of –0.6 at a wavelength of 780 nm, at the very edge of visible red light.

Another advance has come with an advantageous new type of cloak proposed in 2008. This is the carpet cloak, meant to be placed over a curved reflecting surface with an object under it. The cloak makes the surface appear flat and the object is hidden under a "carpet." The design does not require extreme exotic values of n, but only $n > 1$ varying in a specified way over the cloak, and so can be made without the metal split ring resonators of the earlier metamaterials.

In 2009, D. R. Smith and colleagues at Duke University and Southeast University, Nanjing, China showed that a carpet cloak is feasible at microwave frequencies. Their cloak contained thousands of millimeter size H-shaped elements of varying size etched onto a standard copper electronic circuit board, to give values of n from 1.1 to 1.7 properly distributed to provide carpeting. The cloak was completely successful, hiding a bump just as predicted for wavelengths from 1.9 to 2.3 cm (frequencies of 13 to 16 GHz) rather than at one value only.

What's different about carpet cloaks is that they also work for much smaller wavelengths because they can be made at tiny scales from the semiconductor silicon — the basic material for electronic chips that can be manipulated with a highly developed nanotechnology. In 2009, several research groups made silicon carpet cloaks for infrared wavelengths around 1,500 nm, not too distant from the visible range 400–750 nm and typical for optical fiber use. For example, J. H. Lee and colleagues at the University of Colorado, Georgia Institute of Technology, and Hankuk University, Korea, made a successful cloak with a 215 × 80 array of nanometer-size vertical silicon rods, like a miniature orchard, at a regular spacing of 150 nm. Since n was defined by the relative amount of silicon, the rod diameters were controlled during nanofabrication to produce the desired variation in n.

Though these results illustrate carpet cloaking at short wavelengths, all were done in a flat two-dimensional geometry. In 2010, Tolga Ergin of the Karlsruhe Institute of Technology, Germany and colleagues reported the construction of a complex "woodpile" cloak made of nanometer scale plastic rods piled in alternating cross-hatched fashion like logs for a campfire. This hid a bump in a gold film as seen over wide angles, representing for the first time three dimensional cloaking for wavelengths of 1,400 nm to 2,700 nm.

Carpet cloaking has its own problems; theoretical studies have suggested limitations in its ability to completely obscure an object and in the size of the objects it can hide, roughly equal to the wavelength of the illumination. But research on invisibility is rapidly evolving. For instance, in late 2010, two different groups (one from MIT and the SMART Center, Singapore, the other from Imperial College London, the University of Birmingham, and the Technical University of Denmark) announced similar breakthroughs. They had successfully hidden objects up to centimeters in size over visible light wavelengths from blue to red. Remarkably, these encouraging results did not require intricate metamaterials with thousands of elements, but were achieved using the anisotropic optical properties of the naturally occurring crystal calcite.

Invisibility science is only a few years old and results like these make it clear that many surprises are in store. Certainly the work will continue because of the lure of the idea and its scientific challenge; and pragmatically, because it will continue to be funded. Military needs drove the first steps toward invisibility in the form of early passive camouflage, and now continue to drive the adoption of its modern high tech equivalents — active camouflage, stealth technology, and cloaking.

Achievements like working out transformation optics, creating intricate metamaterials, and making aircraft nearly invisible to radar are pushing electromagnetic and photonic science along with nanotechnology toward new limits. They're worth celebrating as steps toward an old dream of invisibility, especially since scientists would have considered the very idea laughable a few years ago. But there's still a long way to go before scientific invisibility replaces the magical kind or the science fictional kind, before it becomes possible to wrap oneself in a cloak or push a button and disappear — if indeed that ever does becomes possible.

Questions of whether the science and the fiction will eventually converge or diverge hover over these and the other aspects of light we've considered, from entangled photons to hydrogen fusion induced by laser. How much of this will remain science fiction? How close are the real and the imaginary science of light? We'll look for answers in the next chapter.

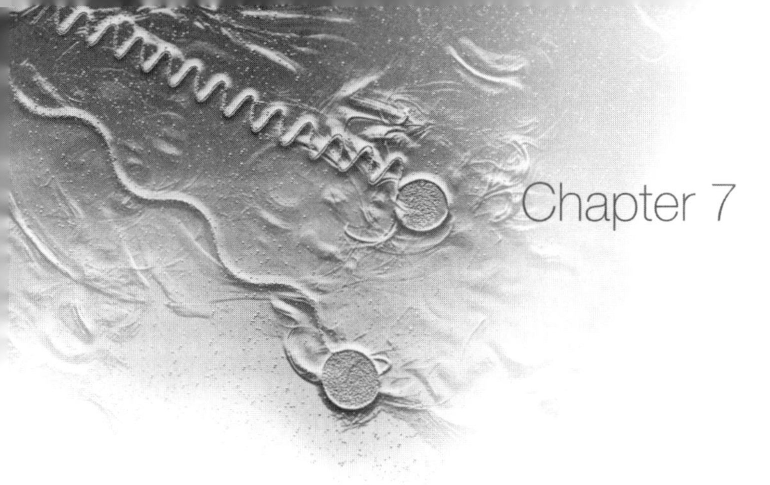

Chapter 7

Light Fantasy to Light Reality

New technologies have traditionally elicited disbelief when they were introduced, probably ever since some early human amazed the rest of the tribe by wielding the first stone axe. Later, steam engines, automobiles, and flying machines were seen as wondrous when they were new. That's in keeping with Arthur C. Clarke's famous dictum, one of his Three Laws of Prediction: "Any sufficiently advanced technology is indistinguishable from magic." The sense of magical awe also accompanied the early electrical technology that led to today's electronic and photonic science. In 1896, 16 years after Thomas Edison patented his incandescent light bulb, one observer of the new field of electric lighting was still sufficiently dazzled to say, "One miracle has followed another until we can only wonder what apparent impossibility will be accomplished next."

But since the mid-20th century, we have become more blasé about succeeding waves of technology. The progressions from radio to big screen high definition television, from vinyl records to MP3 music players, from room-sized computers to smart phones have all been greatly appreciated by consumers eager to snap up the new devices, but without much sense of awe-inspiring magic. That's partly because the intervals between successive advances are becoming ever shorter so there's barely time to digest one wave before the next sweeps in.

The quick acceptance of these modern miracles also owes a great deal to speculations about science and technology embedded in science fiction. These go back to the 19th century musings of Jules Verne and even earlier, but they have gained power in the last decades because the impact of science fiction has grown enormously. In the 1920s and 1930s, science fiction was a fringe genre mostly found in "pulp" magazines of low literary merit. But during and after World War II, developments in areas like nuclear physics, bioscience, and space exploration have made science prominent in the public consciousness and carried science fiction along with the science.

That has been especially true in film. Starting with breakthrough motion pictures like *2001, Close Encounters of the Third Kind* and *Star Wars* in the 1960s and 1970s, and continuing to contemporary blockbusters, the popular and commercial success of science fiction movies has made them a major cultural force. Science fiction television series like *Star Trek* and *Battlestar Galactica* have also had enormous influence, as has the Internet. What's offered on the Internet is often not subject to any fact checking. As a result, speculative or invalid science can be presented on a par with valid science, blurring the boundaries between reality and fantasy. Another factor that softens the distinction is that today's scientific progress can be so rapid that it overtakes fiction!

All this helps explain the deep connections between the properties of light as presented in fantasy fiction, and the real science of light in the 21st century. One other feature binds the real and imaginary science of light, the "eeriness" factor that derives from quantum physics and to a lesser extent from the theory of relativity. For example, entanglement and quantum teleportation are like nothing in ordinary human experience and can't be easily explained. Speculative fiction, however, has presented teleportation and matter transmission since the 19th century, long before "beaming" appeared in the *Star Trek* universe. Now that science is seriously examining entanglement, it's not so obvious where reality ends and fantasy begins.

This blurring shows up in scientists' ready acceptance of the terminology of science fiction and fantasy to describe their work. In 1993, when C. H. Bennett of IBM and colleagues proposed their quantum method of securely sending data over telecommunications channels, they deliberately chose to call it "teleportation" as "a term from science fiction." And would the recent work that uses metamaterials to hide objects be called "cloaking" unless *Star Trek* had used the term — which itself resonates with the magical cloaks that have conferred invisibility on heroes from Siegfried to Harry Potter?

More interesting than terminology is the question of how well science fiction portrays the stranger effects of light, and how accurately it points to their future possibilities and development. In the rest of this chapter, I'll examine how close we are to attaining what fantasy has already attained or predicted, from traveling faster than light, to inducing hydrogen fusion with light, to creating a real cloak of invisibility. Like any kind of forecasting, predictions of where light is going are dangerous but also irresistible, because this important technology shapes our world today and also in the future, from the everyday use of cell phones to the possibilities of exploring our universe. And that's where we'll begin, analyzing our chances of getting out to the rest of the universe in some reasonable time.

Faster than Light Travel

We have to start by recognizing this fact: direct, brute force acceleration of spacecraft to light speed or significant fractions of it is far beyond any foreseeable capabilities using NASA-style rocket engines. The maximum launch speed achieved so far is only 0.005% of c, and weight considerations prevent spacecraft from carrying enough fuel for a sustained push that would produce higher speeds.

There are however other ways to propel spacecraft, including one method that needs no fuel because it uses available light. Since photons carry momentum, they exert a force called radiation pressure when they encounter a surface. Using this, space "sailing ships" could be driven by light from our Sun pushing huge sails made of thin, lightweight material. Though the push would be small, operating over long periods it could provide substantial acceleration. In May 2010,

JAXA — the Japan Aerospace Exploration Agency — launched by conventional means its experimental Ikaros spacecraft (Interplanetary Kite-craft Accelerated by Radiation of the Sun). In July, the craft deployed a sail 200 m^2 in area. This is much smaller than the sails needed for serious acceleration, but the craft did display an additional push due to light. Other experiments are in progress.

It would be satisfying if propulsion by light could be used to approach the speed of light, but solar sailing speeds are not expected to be greater than several times those achievable by conventional engines — still only a tiny percent of *c*. A variant of the idea would use ground-based lasers at gigawatt levels to push the sail hard enough to reach much bigger fractions of *c*. But lasers of this magnitude that can operate continuously rather than in brief pulses are well beyond present technology — and a speed of 0.1 *c*, say, still represents a 40 year voyage to our neighboring star, Proxima Centauri.

Even if these schemes were to produce speeds at large fractions of *c*, special relativity disallows travel faster than *c*, limiting our ability to explore the further universe. But at the moment, there is no serious evidence that special relativity is wrong or that faster-than-light tachyons exist.

General relativity offers other approaches to FTL travel that don't violate special relativity, through a wormhole shortcut or via a warp drive like that proposed by Miguel Alcubierre. Both, however, require cosmic amounts of matter or equivalently energy because they involve radically curved spacetime, and also require that the matter or energy be negative. We know of no negative matter in the universe, but negative energy exists, at least in small amounts. Alcubierre had noted that this exotic condition could be achieved if another aspect of quantum physics, that a vacuum isn't really empty, is considered through the Casimir effect.

In 1948, the Dutch theoretical physicist Henrik Casimir predicted that two metal plates a small distance apart, placed in vacuum, would feel a hitherto unknown attractive force. This force literally arises from "nothing," because in quantum physics, a vacuum is not featureless.

Even with everything removed, vacuum always contains a small amount of electromagnetic energy that fluctuates, according to the statistical nature of quanta, around an average of half the energy of a photon. The resulting electromagnetic fields between two metal plates push them outwards by radiation pressure. But photon wavelengths that don't fit exactly into the gap don't contribute, so the outward push is less than the inward push from fluctuations outside the plates. The net result is that the plates act as if mutually attracted. The region between them represents negative energy, since its energy density is less than that of the "ordinary" vacuum outside the plates, defined as zero energy.

Though the experiments are not easy, the Casimir effect has been shown to exist and so it may not be impossible to achieve the conditions for the Alcubierre drive. Also variants of the drive have been proposed that eliminate some difficulties, though they raise others. But all this remains speculative. In 2007, the British Interplanetary Society presented "Warp Drive, Faster Than Light: Breaking the Interstellar Distance Barrier," a workshop to review the theory and practice of warp drives, including the role of the Casimir effect. Participants noted the inability to build a drive at present, though one projection suggested that one might be built by 2180.

In short, FTL travel is impossible under present knowledge and engineering capability. It seems that the science fiction staples of warp drives and other imaginary approaches will remain a fantasy for well over a century according to the projected date of 2180, if not forever. But the wild card is that a conceptual breakthrough that combines quantum physics and general relativity in a true "Theory of Everything" might give new approaches to FTL. One candidate for this unified approach, string theory, does predict a violation of the constancy of the speed of light; but again, until firm experimental evidence appears, this is only a tantalizing theoretical possibility. Meanwhile, although it may not produce speeds near *c*, the prospect of light-propelled space travel is an appealing science fiction idea (appearing, for instance, in Arthur C. Clarke's story "Sunjammer" from 1964) that could turn into real science.

Faster than Light Communication

Under special relativity, what limits the speed of matter is that it would take infinite energy to reach *c*. But "information" is less tangible than a chunk of matter and harder to define, as expressed in the discussions in Chapters 3 and 4 about phase, group and signal velocities and about what is actually conveyed by fast and backwards light. The arguments against superluminal communication — the violation of causality and the impossibility of time travel — are logical and philosophical rather than physical. Also, unlike material bodies, where reaching even a considerable fraction of *c* would be an achievement, radio communication in free space already occurs at speed *c*.

This seems to give leeway in finding ways to communicate at superluminal speeds, especially with quantum weirdness taken into account. If entanglement does represent some sort of transfer, the measurements made in 2008 that I cited earlier show that it happens at a speed no slower than 10,000 *c*. But so far, in every consideration of methods that use entanglement such as quantum teleportation, a complete message can't be transferred without some auxiliary contact over a conventional channel at the speed of light or less, which defeats the idea of an ansible.

Most physicists would agree that like FTL travel, instantaneous or FTL communication is impossible under present knowledge. But the door to FTL communication is perhaps not completely closed — or at least, were such communication ever shown to be possible, that would be less surprising than FTL travel. In any case, scientists continue to speculate about both possibilities.

Slow, Stopped, Fast, Backwards, and Left-Handed Light

Experiments that slow and stop light, or send its pulses faster than *c* and backwards, enhance the understanding of light and its behavior in exotic systems like Bose–Einstein condensates. These outcomes have made physicists think more carefully about the speed of light and about information transfer, the same issues that arise in FTL communication.

Similar results can now be obtained in room-temperature solids such as optical fibers, which are better suited to real-world use than low temperature atomic BECs. (However, there has been a breakthrough in condensing light itself. Though this is difficult to do, in late 2010 Jan Klaers and colleagues at the University of Bonn made a BEC composed entirely of photons. This "super photon" is expected to function as a novel laser-like light source in new short wavelength regions).

Applications for slow light include optical delay lines and interferometers, optical devices that allow extremely accurate length measurements. These already exist in rudimentary form and can probably be fully developed well within a decade. The application with the greatest potential, though, is the use of stopped light to store data, which would be essential for a quantum information technology based on light. That has been demonstrated only in the laboratory, as has much of quantum information technology, and will develop as that area develops. However, in bringing light to a full stop, science has had the rare distinction of performing a feat more impressive than science fiction's accomplishment of merely slowing light down to 10 m/sec as in *Redshift Rendezvous*. Science may also lead us to a real form of Bob Shaw's slow glass, but only with far more modest storage than ten years compressed into six millimeters.

The other form of recently explored unusual light is left-handed light that travels through a metamaterial with negative refractive index. The most fascinating application of metamaterials is invisibility, which I discuss separately, but the same principles also lead to high resolution superlenses with potentially revolutionary applications in biological science, chip manufacture, and data storage. Since the ideas behind negative *n* and metamaterials can be applied to any kind of wave, not only light, there are also such theoretical possibilities as acoustic superlenses that would provide higher resolutions for the sound waves used in ultrasound scanning.

No one seems to have yet made a general purpose optical device using superlenses, but progress in the laboratory has been rapid. In 2010, N. Fang at the University of Illinois-Urbana, and co-workers at Hewlett-Packard Laboratories, Palo Alto, and the University of

California Davis, used thin, ultra-smooth silver layers to construct a superlens with the very high resolution of wavelength divided by 12. Made with complex but achievable nanotechnology, the lens resolved tiny features only 30 nm across under light of wavelength 380 nm — many times better than a conventional lens and comparing favorably to a theoretical superlens limit of wavelength divided by 20. Though there are basic problems to resolve before a commercial superlens microscope can be made, the payoff is great. It's likely that specialized applications, such as in the semiconductor industry, will appear well within a decade.

Quantum Information Technology with Light

Although the idea of manipulating data through fundamental quantum principles goes back several decades, it has yet to inspire a mature and complete technology. The use of photons as qubits to achieve this is one possibility, along with various atomic, solid state, and superconducting systems that can display two quantum states in superposition to represent binary 1 and 0. Among these, the photonic approach is one of the more developed methods. Photon entanglement has been convincingly demonstrated over large distances, and quantum teleportation of photons for cryptography is commercially available.

The cryptographic application will drive the growth of quantum telecommunications, so the science fiction idea of teleportation has already arrived in the real world. But we're still a long way from laptops that teleport data with quantum chips. In an era where the word "quantum" has become a trendy part of popular science, one might well ask "Where Is My Quantum Computer?" as P. Hemmer and J. Wrachtrup (of Texas A&M University and the University of Stuttgart, respectively) do in a recent review.

As these authors note, there are good reasons to look forward to quantum technology, though it won't necessarily enhance general computation. It's true that the superposed nature of qubits makes it possible to carry out many calculations at once, but not every application can benefit from "quantum parallelism," which will be most useful for certain problems that are simply intractable by conventional means. One is the factorization of large integers, which means determining the prime numbers whose product yields the integer (for instance, $120 = 2 \times 2 \times 2 \times 3 \times 5$, all of which are prime, that is, exactly divisible only by themselves and one). This is important for non-quantum cryptography where extremely large numbers are widely used as encoding keys that can't be broken without finding the prime factors. Conventional computers are so slow at this that they're useless for code breaking. In one example, a team that used special methods to factor a 232 digit number into its two primes estimated that this would take 2,000 years on a standard machine.

But in 1994, the American mathematician Peter Shor showed that factorization would progress exponentially faster on a quantum machine, so this application of quantum methods is important for code breaking, with national security implications. A quantum computer is also superior for searching large databases. Another application that especially excites scientists is the exact simulation of physical systems that are inherently quantum mechanical, as proposed by Richard Feynman in the 1980s. And for the average consumer, Hemmer and Wrachtrup note that the sensitivity of quantum entanglement to eavesdropping, as demonstrated in quantum cryptography, could safeguard against unauthorized collection of data on the Internet.

These benefits motivate the rise of quantum computing, many of whose physical and logical techniques have been tested in photonic quantum cryptography. In the BB84 quantum key protocol I discussed in Chapter 4, if Alice transmits a photon in either the horizontal 0 state or the vertical 1 state, and Bob receives it in the D± basis, that changes 0 or 1 into a superposed state of both 0 and 1. This logical operation, called a Hadamard gate, is a basic single qubit manipulation, and entanglement in quantum cryptography can be described as a two-qubit process. These two types of logical operations are sufficient to build complete programs for quantum computers.

Various researchers have demonstrated quantum feasibility by building minimal two-qubit computers that solve relatively simple problems. Some of these computers use light; others serve as testbeds for alternate approaches. For example, in 2009, L. DiCarlo of Yale University and colleagues from the University of Waterloo, Sherbrooke University, and the Institute of Atomic and Subatomic Physics, Vienna, illustrated a solid state approach by building a two-qubit quantum computer on a chip that employed superconducting electronics.

In 2010, another minimal quantum computer illustrated Feynman's notion of exact simulation by calculating the quantum energy levels of the simplest possible molecule, H_2. Researchers at the University of Queensland in Australia, Harvard, and Truman State University in Missouri used optical methods to logically manipulate polarized photon qubits. After multiple iterations, their results matched the known energy levels of H_2 to within a few parts per million. This is significant because calculating the exact properties of molecules with more than about ten atoms is beyond the capabilities of the biggest conventional computers. Quantum computers with hundreds of qubits could potentially deal with big biomolecules, providing a new tool for biology and for drug design at the molecular level.

But it's challenging to construct computers using hundreds or even tens of qubits. Whether embodied in light or in matter, one fundamental problem is that qubits are delicate. If a quantum system is not isolated from its environment, the interaction can destroy the quantum properties — a qubit that simultaneously holds both 1 and 0 can "decohere" to either 1 or 0. Photonic qubits offer the advantage that they have only limited interactions with their surroundings. On the other hand, generating and sensing multiple single and entangled photons isn't yet a reliable process. Also, using photons to store as well as transmit data is only a recent capability, based on the new techniques of stopped light. It may be that a combination of photonic and atomic systems, as suggested by work that I cited in Chapter 4, will prove the best solution for the manipulation of quantum data.

Far from showing off a shiny new quantum computer, no one can yet predict which approach is most likely to lead to one. However, extensive efforts are under way to determine this, according to an expert panel convened by ARDA (Advanced Research and Development

Activity), a U.S. government agency associated with the intelligence community. The panel's report "A Quantum Information Science and Technology Roadmap" (2004) called the field "rapidly evolving" and "one of the most active research areas of modern science." The report lists 155 research groups, distributed internationally in academic, industrial and government laboratories, and presents benchmarks to reach by 2012. One important goal is to develop error correction methods that compensate for decoherence effects.

Judging by the flood of research papers that continues to appear, funded by a wide variety of sources, activity has only increased since 2004. It's highly likely that next level quantum computers, using more qubits than current versions along with error correction, will be functioning within a decade. That may provide a sufficient basis to construct following generations of truly powerful units for cryptographic and scientific use — and photon technology is certainly in the running. Quantum computation for consumers may never be a viable proposition, but quantum methods that ensure user privacy over the Internet could be a first consumer application, as Hemmer and Wrachtrup suggest, also within a decade.

Laser Fusion

The greatest contrast to the infinitesimal power of a single photon traversing a quantum computer is the unimaginable stream of photons from a really powerful laser. Such lasers exist or have been proposed for uses from weaponry to spacecraft propulsion, but among the biggest, with potentially the biggest impact, are those powerful enough to induce hydrogen fusion.

Science fiction makes laser fusion look easy. In *Spider-Man 2* (2004), four puny laser beams is all it takes to instantly initiate a fusion reaction that glows like a miniature star and then blows up a laboratory. The reality is more demanding, requiring extraordinary laser power and precision, which the National Ignition Facility hopes to deliver using not four but 192 laser beams. These are designed to work in careful synchronization to deliver a total of 1.8 megajoule (MJ) in the form of UV light. That's enough energy to lift a 180 kg mass 1 km straight up, more than any other existing laser system.

To ignite fusion, the beams will be focused on a hohlraum ("cavity," in German), a small gold-plated can the size and shape of a pencil eraser. It contains a spherical fuel capsule a few millimeters across that holds a mixture of the hydrogen isotopes deuterium and tritium. The UV light heats the gold to millions of degrees. This produces X-rays that rapidly compress the hydrogen to over 100 million K at enormous pressure. In theory, that will fuse the hydrogen nuclei into helium and release far more energy than had gone into the reaction. This method is called indirect drive. The advantage over direct drive, where laser energy is delivered straight to the fuel capsule, is that the implosion proceeds more symmetrically and avoids undesirable instabilities. This is essential to reach the temperature within the hydrogen that will induce fusion.

In gearing up for the actual attempt to ignite fusion, NIF passed some important milestones in 2010. Working at a laser energy of 0.7 MJ, NIF researchers demonstrated that the hohlraum reached the necessary temperature and produced a sufficiently symmetric distribution of X-rays to initiate fusion, according to calculations. The NIF also delivered bigger jolts up to 1.3 MJ to implode target fuel capsules, both empty, and loaded with a test mixture of hydrogen fuel for diagnostic purposes. Though these energy levels are less than the system's rated energy of 1.8 MJ, the test shots still represent many multiples of the world's entire electrical power output at any given time, though the laser pulses last only nanoseconds.

These encouraging results are inspiring planning for laser fusion as a practical power source, though in modified form. The U.S. Department of Energy LIFE initiative (laser inertial fusion energy) would build on the presumed success of NIF in inducing fusion to form part of a hybrid fusion-fission scheme. LIFE would use neutrons from laser fusion, an intrinsic byproduct, to extract heat from fissile materials including nuclear waste, which would then generate electricity by conventional means. Assuming that NIF achieves fusion power levels of 20 to 35 MJ in the next couple of years, scientists at Lawrence

Livermore anticipate that a pilot LIFE power plant could be operating by 2020 with a commercial version by 2030, and that a large fraction of U.S. energy needs could be met with a string of LIFE plants by 2050 to 2100.

Laser fusion is one of several approaches that have been tried over six decades of attacking the fusion problem. The most spectacular potential alternative is ITER (originally, the International Thermonuclear Experimental Reactor), under construction at Cadarache in the south of France. This effort is supported by the European Union and six non-European nations including the United States. It will use magnetic confinement, where a mixture of deuterium and tritium is heated to 150 million K in a huge machine called a Tokamak, to form a plasma that will support fusion. Powerful magnetic fields shape the plasma and hold it away from the Tokamak walls. The project, to cost €16 billion, has suffered from rising costs and management problems, with first plasma production not expected until 2019. Other laser fusion projects are also underway. A French equivalent to NIF, Laser Mégajoule, is under construction near Bordeaux, and HiPER (High Power Laser Energy Research) is a proposed European facility that will use a different method than NIF.

Among all these, the NIF leads in the competition to produce fusion, but that still won't come as easily as it did in *Spider-Man 2*. NIF initially projected that ignition would be attained in 2010, but that has now been pushed back to 2011 or 2012. And even if ignition is achieved, it will be necessary to upgrade NIF's lasers to work within the LIFE approach. LIFE also raises questions about handling and storing fissile material; the LIFE scheme is not quite the completely clean power that fusion proponents have dreamt of. The issue is ripe to become politicized, as can also happen to the process of justifying and funding a new energy technology. Also vulnerable to criticism is NIF's parallel support from the National Nuclear Security Administration for the study of thermonuclear weapons, which some observers claim overshadows the fusion part of the program.

It does not diminish NIF's strong scientific achievements to say that at this point it seems premature to predict a pilot power plant in a mere ten years. But it is highly likely that in ten years, after seeing where NIF, LIFE, ITER, and so on take us, we'll have an excellent, perhaps definitive idea of which approach is worth further time, money and effort.

Invisibility

Of the phenomena of light we've considered, invisibility — at least, personal invisibility — may strike the greatest chord because of its aura of power and mystery. That's why it has such a long history in myth and fantasy. Yet invisibility is also the least mysterious in its foundational science, for it doesn't draw on counterintuitive quantum physics or relativity. The three types of invisibility I considered in Chapter 6 — stealth, optical camouflage or virtual transparency, and cloaking with metamaterials — use established, well understood electromagnetic wave theory from the 19th century. The breakthroughs come from modern computational power, materials, and optical technology.

Within its limited scope, the most successful of the three techniques is stealth technology. As it has developed, the decrease in the radar cross section of a large military aircraft to the size of a bird or insect is impressive. But even such a small signature can perhaps be reduced further, and invisibility in another spectral range, the infrared, is also a concern. As military technologies, future details of these developments may not be greatly forthcoming, but these are mature areas where known techniques can be refined.

The optical camouflage that uses retroreflection to produce virtual transparency is striking because it occurs with ordinary light in real life situations, allowing a viewer to apparently see through a person. As the method stands, it could be valuable for certain specific situations. But the question for the future is, can it be turned into a true cloak that would function in the real three dimensional world to hide a person or thing from view in any direction?

"Cloaking" also describes the research done to bend light waves so as to hide objects. In this case, science has been more creative than fiction. Researchers have not attempted the impossible task of bending spacetime with huge masses as required by general relativity and implied in the *Star Trek* cloaking device. Rather, they have designed artificial materials — metamaterials — that control light as desired, or most recently, have found a natural material (calcite) to perform the same function. This approach has shown great success in a short time, but the intriguing question is the same as for virtual transparency: can it lead to a true Harry Potter cloak for personal invisibility?

Both virtual transparency and cloaking by bending light have properties that could contribute to such a seemingly magical cloak. Virtual transparency already operates across the visible spectrum, at human scales, and in a cloak-like retroreflective configuration. Metamaterial engineering offers impressive control of light waves, and a natural material like calcite offers simplicity in manipulating light. But both the virtual and the cloaking methods have serious deficiencies. Virtual transparency needs auxiliary equipment and requires that the observer remain in a specified position. Metamaterial cloaking is still difficult to extend to broad spectral ranges, and cloaking with either artificial or natural materials has yet to be demonstrated for meter-size subjects. Indeed, it's hard to see how any kind of material could be fabricated or formed to operate at that scale.

So at present, neither technique could immediately produce a usable personal cloak. But as I suggested in Chapter 6, one can imagine extending the optical camouflage approach into a garment whose back is coated with sensors that relay what they detect to devices on the front that display those images to create virtual transparency. In their 2002 paper "The Cloak of Invisibility: Challenges and Applications," Franco Zambonelli and Marco Mamei, of the University of Modena and Reggio Emilia, Italy, examine this idea in detail and reach quantitative conclusions about the necessary hardware and software.

The authors begin by defining their basic cloaking system: "a fabric of small computing devices that can receive and retransmit light emissions in a directional way as well as interact with each other in a wireless amorphous network" in such a manner that an observer sees the same visual scene as if the cloak weren't there. They calculate parameters for the receivers and transmitters, called In and Out types respectively, and for the supporting computational power, for three cases of escalating complexity: a rigid wall; a rigid object of arbitrary shape; and a true Harry Potter style cloak that changes as needed to follow the motion of an object or person hidden under it.

Zambonelli and Mamei estimate that to make a 1 m^2 wall invisible from 10 m away, at a resolution that looks natural, would require 116,000 Ins (such as CCD sensors for digital cameras) on the back of the wall and an equal number of Outs (such as LCD displays or LEDS) on its front. To correctly display the scene behind the wall on its front, these devices would need to form a network communicating their locations and visual data via short-range wireless. These considerations translate into device sizes of 8 mm^2 and a data handling rate of 300 kbits/sec. Both, the authors write, are achievable with existing technology, that is, the technology of 2002.

The requirements go up sharply, however, for the more complicated cases. To render invisible a rigid sphere 1 m in diameter, as seen from any direction, it would need to be covered with tiny devices less than 5 μm across, supported by a network with a data handling rate up to 1 Tbit/sec. For a flexible cloak whose conformation can change, even more computational power and bandwidth would be needed to constantly recalculate the spatial relations among the In and Out devices, an extremely demanding requirement. On the positive side, the movements of the cloak could be converted into electrical power for the device network. Also the cloak might not be too exorbitantly expensive. Based on a projection of the future cost of very small devices, Zambonelli and Mamei estimate that a cloak of area 3 m^2 would cost under €500,000.

As the authors note, the device specifications for the flexible cloak were barely achievable in 2002. Now, a decade later, nanotechnology has advanced and suitable devices are within reach. For instance, the quantum dots discussed in Chapter 5 — artificial atoms a few nanometers across — can serve both as light detectors and sources, that is, as Ins and Outs, at the necessary small scales. The dots can be formed in large arrays by various techniques, one of which seems well adapted to covering a cloak with myriad devices. The procedure is to mix the dots into an ink-like solution, and then spray them out in any desired pattern through an inkjet printer. This is typically done onto a rigid semiconductor substrate; but the dots can also be sprayed onto paper and even cotton cloth, as recently reported by Aaron Small of the Victoria University of Wellington, New Zealand, and colleagues.

This is only one possibility within the active research area that seeks ways to deposit small electronic and photonic devices on thin, flexible surfaces for a variety of applications. So an actual cloak configuration seems possible, though there remain big problems of device performance as well as the networking and computational capacity needed for a sufficiently rapid cloaking display. Nevertheless, electronic and photonic development is proceeding quickly. With sufficient funding, perhaps from the military, the first steps described by Zambonelli and Mamei — making a wall or rigid object invisible — could be reached within a decade.

But apart from the traditional invisibility to the eye celebrated in fantasy, there's another kind that may have greater impact. Now that scientists know how to make metamaterials that guide light waves around an object, they are applying the idea to situations where avoidance of a given area could save lives and property. Along with Harry Potter invisibility cloaks, we may look forward to large scale cloaks that divert earthquake waves around buildings and prevent ocean waves from damaging coastal structures — a kind of invisibility that did not appear in the myth and magic of cloaking and represents science creatively extending fantasy in unexpected directions.

Epilogue

We began this examination of light with an imaginary tour through the universe to point out light's variety and prevalence, as well as its strange characteristics. If you are impressed by light's importance, but still puzzled by its apparent contradictions and impossibilities, you have good company in Albert Einstein. He knew light well — after all, it was Einstein who gave the speed of light its central role and brought us the photon — but in rejecting the quantum world that the photon represents, he also represented contradiction. Einstein once said that he would reflect on light for the rest of his life, but like every researcher and thinker who has examined light, he never completely penetrated its mysteries. Some of those mysteries extend beyond the science of light, such as how the Higgs boson explains the zero mass of the photon along with the masses of all the elementary particles, and the connections between quantum physics and general relativity, which may affect our understanding of light.

The other explorers of light are science fiction writers and filmmakers, who have more freedom to speculate than scientists. The speculation may be disconnected from reality, but it can also extend existing science in meaningful directions. Among Arthur C. Clarke's Rules of Prediction is this: "The only way of discovering the limits of the possible is to venture a little way past them into the impossible." In many cases, that is what science fiction has done. Sometimes the science cannot accommodate even the first step into the impossible, as with FTL travel; but just look at the research in teleportation or invisibility, and you see imagination turning into reality.

One could argue that there is still a great gap in knowledge, that by applying quantum effects we don't fully understand, we're omitting something important. Perhaps so: but it's also true that merely working with the technology enlarges understanding. By using teleportation, for instance, scientists and engineers begin to develop intuitions about photons and quanta in general. The generation of researchers who take teleportation for granted within both science fiction and an everyday technology of light may be the generation that, in theorist John Wheeler's memorable words, discovers the "utterly simple idea that demands the quantum."

That would finally resolve the mysteries of quantum mechanics and make the bizarre aspects of light far less bizarre. But by that time, imaginative science fiction writers will have pushed the boundaries of the possible into new mysteries for creative scientists to translate into reality, as science and science fiction continue to draw on each other for their mutual benefit and for the benefit of humanity.

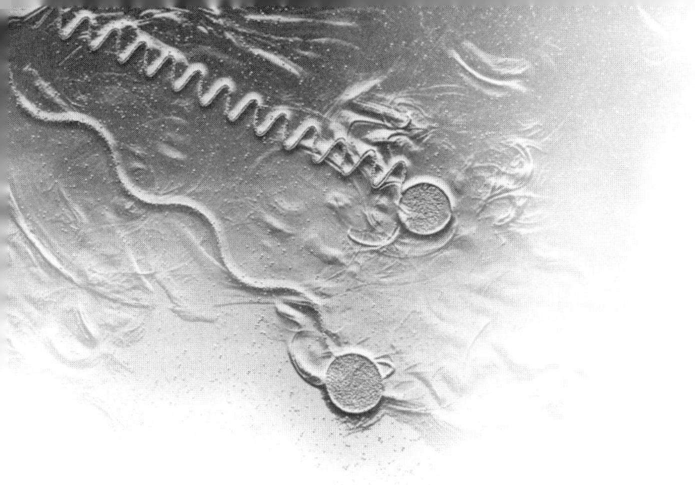

Acknowledgements

It's a great pleasure to thank my agent, Laura Wood, of Fineprint Literary Management, for her immediate appreciation of this book idea. Thanks also to David Parsons, my editor at World Scientific United States, who also encouraged the idea. Both Laura and David invested considerable time and effort to ensure that the book proceeded as planned. Without their joint labors, this book would not have happened. Thanks go as well to Alvin Chong in Singapore and the production staff at World Scientific India, who efficiently edited and processed the manuscript.

At Emory University, undergraduate Robert Lunde worked hard and ingeniously to help research the book in an accurate and timely manner. Marc Merlin, of the Atlanta Science Tavern, drew on his undergraduate training in physics at Emory and his graduate research experience in elementary particle physics to make valuable comments on the manuscript.

Across town at Georgia Tech, my colleague Lisa Yaszek fielded questions about science fiction and its history through her own knowledge and by introducing me to the Science Fiction Research Association. Several members of the group took the trouble to answer my queries: John Pierce, Nathan Rockwood, A. William Pett, Hal W. Hall, Edward James, Greg Tidwell, Andy Sawyer, Sha Labare, and Ritchie Calvin.

Needless to say, although every author lives in the dream that his or her book will be completely error free, any faults lie with me, not with my valued contributors.

A different contribution came from Tom and Laurie, proprietors of the Bald Eagle Coffee Shop in Cannon Beach, Oregon, who provided a welcoming environment, good coffee, and reliable WiFi as I worked through the final editing of this book.

My deep thanks to you all.

Sidney Perkowitz
Atlanta, Georgia and Cannon Beach, Oregon
May–December, 2010

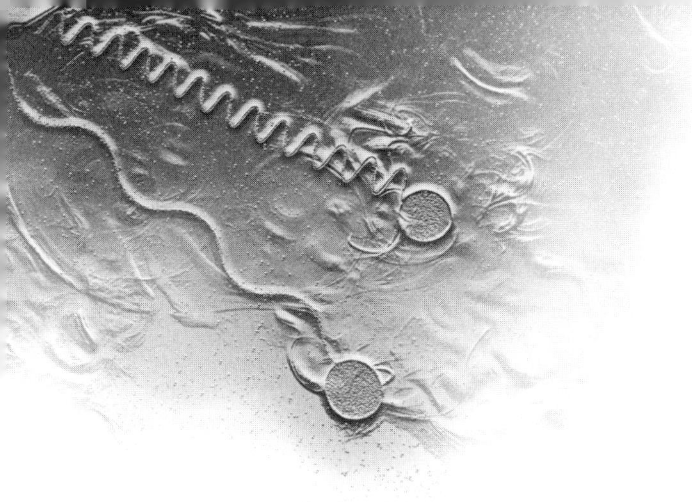

Further Reading, Surfing, and Viewing

Printed works, websites, films and television shows cited in this book or that extend its topics including popular and historical coverage; fictional treatments; and for those who want more detailed science, research articles in scientific journals. Film entries give the director and year of theatrical release.

Chapter 1 What is Light? The Mystery Continues

Background and popular treatments

Einstein, Albert. *Relativity: The Special and the General Theory.* New York: Wings Books, 1961.

Lindley, David. *Where Does The Weirdness Go?* New York: Basic Books, 1997.

Park, David. *The How and the Why.* Princeton: Princeton University Press, 1990.

———. *The Fire Within the Eye.* Princeton: Princeton University Press, 1997.

Perkowitz, Sidney. *Empire of Light.* Washington, DC: Joseph Henry Press, 1998.

Chapter 2 Why is Light so Fast?

Background and popular treatments

Bobis, Laurence and James Lequeux. "Cassini, Rømer, and the Velocity of Light." *J. Astronomical History and Heritage* **11**, 97–105 (2008). http://www.bibli.obspm.fr/Bobis%20and%20Lequeux.pdf.

Einstein, Albert. *Relativity: The Special and the General Theory.* New York: Wings Books, 1961.

Flammarion, Camille. *Lumen.* New York: Dodd, Mead and Company, 1897. http://books.eserver.org/fiction/lumen/conv1-2.html.

Fowler, Michael. "The Speed of Light." University of Virginia Physics Department. http://galileoandeinstein.physics.virginia.edu/lectures/spedlite.html. Accessed 10/22/2010.

Park, David. *The How and the Why.* Princeton: Princeton University Press, 1990.

———. *The Fire Within the Eye.* Princeton: Princeton University Press, 1997.

Perkowitz, Sidney. *Empire of Light.* Washington, DC: Joseph Henry Press, 1998.

For readers who want more detail

Einstein, Albert. "On the Electrodynamics of Moving Bodies (Zur Electrodynamik bewegter Korper)." *Ann. Phys.* **17**, 891 (1905).

Fischbach, E. *et al.* "New geomagnetic limits on the photon mass and on long-range forces coexisting with electromagnetism." *Phys. Rev. Lett.* **73**, 514–517 (1994).

Lesgourgues, Julien. "Galaxies weigh in on neutrinos." *Physics* **3**, 57 (2010).

Thomas, Shaun A. *et al.* "Upper Bound of 0.28 eV on Neutrino Masses from the Largest Photometric Redshift Survey." *Phys. Rev. Lett.* **105**, 31301 (2010).

Chapter 3 Can Anything Go Even Faster?

Background and popular treatments

Albert, David Z. and Rivka Galchen. "Was Einstein Wrong? A Quantum Threat to Special Relativity." *Scientific American.* Mar. 2009, 32–39.

Einstein, Albert. *Relativity: The Special and the General Theory.* New York: Wings Books, 1961.

Feinberg, Gerald. "Particles That Go Faster Than Light." *Scientific American.* Feb. 1970, 69–77.

Herbert, Nick. *Faster Than Light: Superluminal Loopholes in Physics.* New York: Dutton, 1989.

Millis, Marc G. *Breakthrough Propulsion Physics Project: Project Management Methods*. NASA. http://gltrs.grc.nasa.gov/reports/2004/TM-2004-213406.pdf. Dec. 2004. Accessed 10/25/2010.

Minkowski, Hermann. "Space and Time" in Hendrik A. Lorentz, Albert Einstein, Hermann Minkowski, and Hermann Weyl, *The Principle of Relativity: A Collection of Original Memoirs on the Special and General Theory of Relativity*. New York: Dover, 1952.

Perkowitz, Sidney. "Castles in the Air." *Physics World*, January 2009, 2–5.

Scott, Jeff. "Spacecraft Speed Records." Aerospaceweb.org. http://www.aerospaceweb.org/question/spacecraft/q0260.shtml. Feb. 5, 2006. Accessed 10/24/2010.

Thorne, Kip S. *Black Holes and Time Warps: Einstein's Outrageous Legacy*. New York: W. W. Norton, 1994.

Wheeler, John Archibald and Kenneth Williams. *Geons, Black Holes, and Quantum Foam*. New York: W. W. Norton, 2000.

Fictional treatments

Asimov, Isaac. *Foundation's Edge*. New York: Ballantine Books, 1982.

Benford, Gregory. *Timescape*. New York: Bantam Books, 1992.

Campbell, John W. *Islands of Space*. New York: Ace Books, 1956. (Reprint. Original publication date, 1930). http://www.gutenberg.org/ebooks/20988.

Clute, John and Peter Nichols. *The Encyclopedia of Science Fiction*. New York: St. Martin's Press, 1995. Articles, "space warp," "faster than light."

K-Pax (Film, Iain Softley, 2001).

Moore, Alan and Dave Gibbons. *Watchmen*. New York: DC Comics, 2005. (Film version, *Watchmen*, Zach Snyder, 2009).

Sagan, Carl. *Contact*. New York: Pocket Books, 1985. (Film version, *Contact*, Robert Zemeckis, 1997).

Smith, Edward E. *The Lensman Series*. Baltimore: Old Earth Books, 1997.

Stargate SG-1 (Television series, 1997–2007).

Star Trek (Television series, 1966–2005; film series, 1979–2009).

Sternbach, Rick and Michael Okuda. *Star Trek: The Next Generation Technical Manual*. New York: Pocket Books, 1991.

Stith, John E. *Redshift Rendezvous*. New York: Ace Books, 1990.

For readers who want more detail

Alcubierre, Miguel. "The warp drive: hyper-fast travel within general relativity." *Class. Quant. Gravity* **11**, L73–L77 (1994).

Bilaniuk, Olexa-Myron, P. A. Sudarshan, and E. C. George. "Particles beyond the Light Barrier." *Physics Today* **22**, 43–51 (1969).

Bilaniuk, Oleksa-Myron. "Tachyons," in R. M. Walser, A. P. Valanju and P. M. Valanju, *Sudarshan: Seven Science Quests, 6–7 Nov. 2006, Austin, Texas, USA. J. Phys. Conf. Series* **196**, 2009. http://iopscience.iop.org/1742-6596/196/1.

Einstein, Albert. "On the Electrodynamics of Moving Bodies (Zur Electrodynamik bewegter Korper)." *Ann. Phys.* **17**, 891 (1905).

Farrell, Daniel J. and Jeremy Dunning-Davies. "The Constancy, or Otherwise, of the Speed of Light," in *New research on astrophysics, neutron stars and galaxy clusters.* Ed. Louis V. Ross. New York: Nova Science Publishers, 2007, 67–85.

Feinberg, Gerald. "Possibility of Faster-Than-Light Particles." *Phys. Rev.* **159**, 1089–1105 (1967).

Gisler, Galen. "A galactic speed record." *Nature* **371**, 18 (1994).

Morris, Michael S., Kip S. Thorne and Ulvi Yurtsever. "Wormholes, time machines, and the weak energy condition." *Phys. Rev. Lett.* **61**, 1446–1449 (1988).

Morris, Michael S. and Kip S. Thorne. "Wormholes in spacetime and their use for interstellar travel: A tool for teaching general relativity." *Am. J. Phys.* **56**, 395–412 (1988).

Shapiro, Irwin. "Fourth Test of General Relativity." *Phys. Rev. Lett.* **13**, 789–791 (1964).

——— *et al.* "Fourth Test of General Relativity: Preliminary Results." *Phys. Rev. Lett.* **20**, 1265–1269 (1968).

Uzan, J. P. "The fundamental constants and their variation: observational status and theoretical motivations." *Rev. Mod. Phys.* **74**, 403 (2003).

Chapter 4 Slow, Stopped, Fast, and Backwards Light

Background and popular treatments

Browne, Malcolm W. "Researchers Slow Speed of Light To the Pace of a Sunday Driver." *The New York Times*, February 18, 1999. http://www.nytimes.com/1999/02/18/us/researchers-slow-speed-of-light-to-the-pace-of-a-sunday-driver.html?scp=5&sq=lene+hau&st=nyt.

Holloway, Marguerite. "What Visions in the Dark of Light." *Scientific American* **297**, 50–53 (2007).

Johnston, Hamish. "Slowed light breaks record." *Physics World.* http://physicsworld.com/cws/article/news/41246 December, 2009. Accessed 10/28/2010.

Petit, Charles. "Quantum Computer Simulates Hydrogen Molecule Just Right." *Science News* Jan 28, 2010. http://www.wired.com/wiredscience/2010/01/quantum-computer-hydrogen-simulation/. Accessed 9/25/2010.

Fictional treatments

De Camp, L. Sprague. "The Exalted." *Astounding Science Fiction*, November 1940. http://www.andrew-may.com/asf/list.htm, http://oglethorpe.edu/faculty/~m_rulison/Honors/SpeculativeFiction/Documents/L.%20Sprague%20de%20Camp%20-%20The%20Exalted.pdf.

Pratchett, Terry. *The Colour of Magic*. New York: St. Martin's Press, 1983. (The first of approximately three dozen *Ringworld* novels).

Shaw, Bob. "Light of Other Days," in Brian Aldiss and Harry Harrison, *Nebula Award Stories No. 2*. London: Gollancz, 1967.

———. *Other Days, Other Eyes*. London: Gollancz, 1972.

Stith, John E. *Redshift Rendezvous*. New York: Ace Books, 1990.

Technovelgy.com. "Slow glass rod." http://www.technovelgy.com/ct/content.asp?Bnum=1502. n. d. Accessed 10/25/2010/.

For readers who want more detail

Boyd, R. W. and Daniel J. Gauthier. "Controlling the Velocity of Light Pulses" *Science* **326**, 1074–1077 (2009).

Brillouin, Léon. *Wave Propagation and Group Velocity*. New York: Academic Press, 1960.

Camacho, Ryan M. *et al*. "All-Optical Delay of Images using Slow Light." *Phys. Rev. Lett.* **98**, 43902 (2007).

Choi, K. S. *et al*. "Mapping photonic entanglement into and out of a quantum memory." *Nature* **452**, 67–71 (2008).

Chu, S. and S. Wong. "Linear Pulse Propagation in an Absorbing Medium." *Phys. Rev. Lett.* **48**, 738 (1982).

Garret, C. G. B. and D. E. McCumber. "Propagation of a Gaussian Light Pulse through an Anomalous Dispersion Medium." *Phys. Rev. A* **1**, 305–313 (1970).

Gehring, George M. *et al*. "Observation of Backward Pulse Propagation Through a Medium with a Negative Group Velocity." *Science* **312**, 895–897 (2006).

Hau, Lene Vestergaard *et al*. "Light speed reduction to 17 metres per second in an ultracold atomic gas." *Nature* **397**, 594–598 (1999).

Hau, Lene Vestergaard. "Tangled memories." *Nature* **452**, 37–38 (2008).

Khurgin, Jacob B. and Rodney S. Tucker. Eds. *Slow Light: Science and Applications.* Boca Raton, FL: CRC Press, 2009.

Liu, Chien *et al.* "Observation of coherent optical information storage in an atomic medium using halted light pulses." *Nature* **409**, 490 (2001).

Milonni, P. W. *Fast Light, Slow Light and Left-Handed Light.* Bristol; Philadelphia: Institute of Physics, 2005.

Schweinsberg, A. *et al.* "Observation of superluminal and slow light propagation in erbium-doped optical fiber." *Europhys. Lett.* **73**, 218–224 (2006).

Shahriar, Selim M. and Philip R. Hemmer. Eds. *Advances in Slow and Fast Light III. 25–26 January 2010. San Francisco, California.* Proceedings of SPIE **7612**. Bellingham, Washington: SPIE, 2010.

Stenner, Michael D., Daniel J. Gauthier and Mark A. Neifeld. "The speed of information in a 'fast-light' optical medium." *Nature* **425**, 695–698 (2003).

Wang, L. J. "Gain-assisted superluminal light Propagation." *Nature* **406**, 277–279 (2000).

Zhang, Rui, Sean R. Garner, and Lene Vestergaard Hau. "Creation of Long-Term Coherent Optical Memory via Controlled Nonlinear Interactions in Bose-Einstein Condensates." *Phys. Rev. Lett.* **103**, 233602 (2009).

Chapter 5 Extreme and Entangled Light

Background and popular treatments

Aglibut, Andrea. "Bank Transfer via Quantum Cryptography Based on Entangled Photons." *Universität Wien, Institut für Experimentalphysik.* 21 April 2004. http://www.secoqc.net/downloads/pressrelease/Banktransfer_english.pdf. Accessed 10/24/2010.

Clegg, Brian. *The God Effect: Quantum Entanglement, Science's Strangest Phenomenon.* New York: St. Martin's Press, 2009.

Einstein, A. *The Born-Einstein Letters; Correspondence between Albert Einstein and Max and Hedwig Born from 1916 to 1955,* New York: Walker, 1971.

Feynman, R. P., R. B. Leighton, and M. Sands, *The Feynman Lectures on Physics.* Reading, MA: Addison Wesley, 1964.

Gilder, Louisa. *The Age of Entanglement: When Quantum Physics Was Reborn.* New York: Knopf, 2008.

Lawrence Livermore National Laboratory. "National Ignition Facility & Photon Science." Lawrence Livermore National Laboratory. https://lasers.llnl.gov/n.d. Accessed 10/26/2010.

Lindley, David. *Where Does The Weirdness Go?* New York: Basic Books, 1997.

Pease, Roland. " 'Unbreakable' encryption unveiled." *BBC News*. http://news.bbc.co.uk/2/hi/science/nature/7661311.stm. 9 October 2008. Accessed 10/26/2010.

Perkowitz, Sidney. "From Ray Guns to Blu-Ray." *Physics World*, May 2010, 16–20.

Zeilinger, Anton. *Dance of the Photons*. New York: Farrar, Strauss and Giroux, 2010.

Fictional treatments

Card, Orson Scott. *Ender's game*. New York: Tor, 1994.

Chain Reaction (Film, Andrew Davis, 1996).

Clute, John and Nichols, Peter. *The Encyclopedia of Science Fiction*. New York: St. Martin's Press, 1995. Articles, "matter transmission," "Star Trek."

The Day the Earth Stood Still (Film, Robert Wise, 1951).

Le Guin, Ursula K. *Rocannon's world*. New York: Garland Publishing, 1975 (c1966).

———. *The Dispossessed*. New York: Harper & Row, 1974

Moon, Elizabeth. *Winning Colors*. Riverdale, NY: Baen, 2000.

———. *Command Decision*. New York: Del Ray Books, 2008.

Real Genius (Film, Martha Coolidge, 1985).

RoboCop (Film, Paul Verhoeven, 1987).

Spider-Man 2 (Film, Sam Raimi, 2004).

Star Wars Episode IV: A New Hope (Film, George Lucas, 1977).

Sternbach, Rick and Okuda, Michael. *Star Trek: The Next Generation Technical Manual*. New York: Pocket, 1991.

Wells, H. G. *The Time Machine, The Invisible Man, The War of the Worlds*. New York: Random House, 2010. (Film versions, *The War of the Worlds*, Byron Haskin, 1953; *War of the Worlds*, Steven Spielberg, 2005).

For readers who want more detail

Aspect, Alain, Philippe Grangier and Gerard Roger. "Experimental Realization of Einstein-Podolsky–Rosen–Bohm *Gedankenexperiment*: A New Violation of Bell's Inequalities." *Phys. Rev. Lett.* **49**, 91–94 (1982).

Bell, J. S. "On the Einstein–Podolsky–Rosen Paradox." *Physics* **1**, 195–200 (1964).

Bennett, C. H. and G. Brassard. "Quantum cryptography: Public-key distribution and coin tossing," in *Proceedings of IEEE International Conference on Computers, Systems and Signal Processing, Bangalore, India, December 1984*, 175–179.

Bennett, Charles H. *et al.* "Teleporting an Unknown Quantum State via Dual Classical and Einstein–Podolsky–Rosen Channels." *Phys. Rev.* **70**, 1895–1899 (1993).

Boschi, D. *et al.* "Experimental Realization of Teleporting an Unknown Pure Quantum State via Dual Classical and Einstein–Podolsky–Rosen Channels." *Phys. Rev. Lett.* **80**, 1121–1125 (1998).

Bouwmeester, Dick *et al.* "Experimental Quantum Teleportation." *Nature* **390**, 575–579 (1997).

Buller, G. S. and R. J. Collins. "Single-photon generation and detection." *Meas. Sci. Technol.* **21**, 012002 (2010).

Dixon, A. R. *et al.* "Gigahertz decoy quantum key distribution with 1 Mbit/s secure key rate." *Opt. Express* **16**, 18790–18979 (2008).

Einstein, A., B. Podolsky and N. Rosen. "Can Quantum-Mechanical Description of Physical Reality Be Considered Complete?" *Phys. Rev.* **47**, 777–780 (1935).

Flagg, Edward B. *et al.* "Interference of Single Photons from Two Separate Semiconductor Quantum Dots." *Phys. Rev. Lett.* **104**, 137401 (2010).

Grangier, P., G. Roger and A. Aspect. "Experimental Evidence for a Photon Anticorrelation Effect on a Beam Splitter: A New Light on Single-Photon Interferences." *Europhys. Lett.* **1**, 173–179 (1986).

Haynam, C. A. *et al.* "National Ignition Facility laser performance status." *Appl. Opt.* **46**, 3276–3303 (2007).

Jin, Xian-Min *et al.* "Experimental free-space quantum teleportation." *Nature Photonics* **4**, 376–381 (2010).

Lettow, R. *et al.* "Quantum Interference of Tunably Indistinguishable Photons from Remote Organic Molecules." *Phys. Rev. Lett.* **104**, 123605 (2010).

Marcikic, Ivan *et al.* "Long-distance teleportation of qubits at telecommunication wavelengths." *Nature* **421**, 509–513 (2003).

Patel, Raj B. *et al.* "Two-photon interference of the emission from electrically tunable remote quantum dots." *Nature Photonics* **4**, 632 (2010).

Salart, Daniel *et al.* "Testing the Speed of 'Spooky Action at a Distance.'" *Nature* **454**, 861–864 (2008).

Scarani, Valerio *et al.* "The security of practical quantum key distribution." *Rev. Mod. Phys.* **81**, 1301–1350 (2009).

Schrödinger, E. "Discussion of Probability Relations Between Separated Systems." *Proc. Cambridge Phil. Soc.* **31**, 555–563 (1935); **32**, 446–451 (1936).

Taylor, G. I. "Interference Fringes with Feeble Light." *Proc. Cambridge Phil. Soc.* **15**, 114–115 (1909).

Ursin, Rupert *et al.* "Quantum Teleportation Link across the Danube." *Nature* **430**, 849 (2004).

——— *et al.* "Entanglement-based quantum communication over 144 km." *Nature Physics* **3**, 481–486 (2007).

Villoresi, P. *et al.* "Experimental verification of the feasibility of a quantum channel between space and Earth." *New J. Phys.* **10**, 12 (2008).

Wooters, W. K. and W. H. Zurek. "A Single Photon Cannot be Cloned." *Nature* **299**, 802 (1982).

Zeilinger, Anton. "Entanglement-Based Quantum Cryptography and Quantum Communication." *Bull. Am. Phys. Soc.* **51**, Abstract BAPS.2007.MAR.U2.1 (2007).

Chapter 6 Invisibility

Background and popular treatments

Behrens, Roy R. "The Theories of Abbott H. Thayer: Father of Camouflage." *Leonardo* **21**, 291-296 (1988).

Belfiore, Michael. *The Department of Mad Scientists.* Washington, DC: Smithsonian Books, 2009.

Brooke, James. "Tokyo Journal; Behold, the Invisible Man, if Not Seeing Is Believing." *The New York Times,* March 27, 2003.

Castelvecchi, Davide. "Closer to vanishing." *Science News* **171**, 180 (2007).

Cho, Adrian. "High-Tech Materials Could Render Objects Invisible." *Science* **312**, 1120 (2006).

Day, Dwyane A. "Stealth Technology." *U.S. Centennial of Flight Commission.* http://www.centennialofflight.gov/essay/Evolution_of_Technology/Stealth_tech/Tech18.htm. n.d. Accessed 6/8/2010.

Fountain, Henry. "Strides in Materials, but No Invisibility Cloak." *The New York Times.* http://www.nytimes.com/2010/11/09/science/09meta.html?_r=1&scp=1&sq=ulf&st=cse. November 9, 2010. Accessed 11/10/2010/.

McKee, Kent W. and David W. Tack. "Active Camouflage for Infantry Headwear Applications." Defence Research and Development Canada. http://pubs.drdc-rddc.gc.ca/BASIS/pcandid/www/engpub/DDW?W%3DKEYWORDS+INC+'casques'+ORDER+BY+Repdate/Descend%26M%3D4%26K%3D531626%26U%3D1. 1 Feb 2007. Accessed 10/25/2010.

NTDTV. "Invisibility Breakthrough for Japanese Researchers." http://www.youtube.com/watch?v=PD83dqSfC0Y&feature=related. n.d. Accessed 10/28/2010.

Rich, Ben R. and Leo Janos. *Skunk Works.* Boston: Little, Brown, 1994.

Schowengerdt, Richard N. and Lev I. Berger. "Innovations In Electro-Optical Camouflage: Project Chameleo." *Proceedings of 2005 MSS Parallel Symposium,* 14–18 Feb 2005. http://www.chameleo.net/InnEOCam-PC-Final.ppt. Accessed 10/28/2010.

Smith, Lewis. "A real invisibility cloak? Wizard!" *The Times* (London) http://entertainment.timesonline.co.uk/tol/arts_and_entertainment/article606923.ece. October 20, 2006. Accessed 3/19/2011.

Tucker, Patrick. "Finding Invisibility." *Futurist* **41**, Sep/Oct 2007, 14–15.

Fictional treatments

Dalton, James. *The Invisible Gentleman.* London: Edward Bull, 1833. http://www.archive.org/details/invisiblegentlem01dalt.

Dick, Philip K. *A Scanner Darkly.* New York: Vintage, 1991.

Dr. Strangelove (Film, Stanley Kubrick, 1964).

Jimenez, Phil. "Wonder Woman's Invisible Jet" in Scott Beatty *et al. The DC Comics Encyclopedia.* London: Dorling Kindersley, 2008. pp. 34–35.

London, Jack. *The Science Fiction of Jack London.* Boston: Gregg Press, 1975.

Mitchell, Edward Page. *The Crystal Man.* New York: Doubleday, 1973.

The Philadelphia Experiment (Film, Stewart Raffill, 1984).

Predator (Film, John McTiernan, 1987).

Plato. *The Republic of Plato* (Trans. Benjamin Jowett). (Available from Google books and numerous other online sites).

Rowling, J. K. *Harry Potter Boxed Set.* New York: A. A. Levine Books, 2009. (*Harry Potter and the Sorcerer's Stone* and six others).

Sherman, David and Dan Cragg. *First to Fight.* New York: Del Rey, 1997.

Star Trek (Television series, 1966–2005; film series, 1979 – 2009).

Sternbach, Rick and Michael Okuda. *Star Trek: The Next Generation Technical Manual.* New York: Pocket, 1991.

Tolkien, J. R. R. *The Hobbit, The Lord of the Rings.* New York: Del Rey, 1986.

Wells, H. G. *The Time Machine, The Invisible Man, The War of the Worlds.* New York: Random House, 2010. (Film version, *The Invisible Man,* James Whale, 1933).

For readers who want more detail

Cai, Wenshan and Vladimir Shalaev. "Optical Metamaterials: Fundamentals and Applications." New York; London: Springer, 2009.

Chen, Xianzhong *et al.* "Macroscopic Invisibility Cloaking of Visible Light." ArXiv.org. http://arxiv.org/abs/1012.2783. Dec. 13, 2010. Accessed 12/29/2010.

Dolling, G. *et al.* "Negative-index metamaterial at 780 nm wavelength." *Optics Letters* **32**, 53–55 (2007).

Ergin, Tolga *et al.* "Three-Dimensional Invisibility Cloak at Optical Wavelengths." *Science* **328**, 337–339 (2010).

Fang, N. *et al.* "Sub-Diffraction-Limited Optical Imaging with a Silver Superlens." *Science* **308**, 534 (2005).

Gabrielli, Lucas *et al.* "Silicon nanostructure cloak operating at optical frequencies." *Nature Photonics* **3**, 461–463 (2009).

Inami, Masahiko, Naoki Kawakami and Susumu Tachi. "Optical Camouflage Using Retro-reflective Projection Technology," in *Proceedings of the Second IEEE and ACM International Symposium on Mixed and Augmented Reality (ISMAR '03)*, 2003. p. 348.

Leonhardt, Ulf. "Optical Conformal Mapping." *Science* **312**, 1777–1780 (2006).

——— and Thomas Philbin. *Geometry and Light.* Mineola, NY; Dover Books, 2010.

Lee, J. H. *et al.* "Direct visualization of optical frequency invisibility cloak based on silicon nanorod array." *Optics Express* **17**, 12922–12928 (2009).

Li, Jensen. and J. B. Pendry. "Hiding under the carpet: A new strategy for cloaking." *Phys. Rev. Lett.* **101**, 203901 (2008).

Liu, R. *et al.* "Broadband Ground Plane Cloak." *Science* **323**, 366–369 (2009).

Pendry, John B. "Negative Refraction Makes a Perfect Lens." *Phys. Rev. Lett.* **85**, 3966–3969 (2000).

——— and David R. Smith. "Reversing Light: Negative Refraction." *Physics Today* **57**, 37–43 (2004).

——— *et al.* "Controlling Electromagnetic Fields," *Science* **312**, 1780–1782 (2006).

Schurig, D. *et al.* "Metamaterial Electromagnetic Cloak at Microwave Frequencies." *Science* **314**, 977–980 (2006).

Shelby, R. A., D. R. Smith and S. Schultz. "Experimental Verification of a Negative Index of Refraction." *Science* **292**, 77–79 (2001).

Shelby, R. A. *et al.* "Microwave transmission through a two-dimensional, isotropic, left-handed metamaterial." *Appl. Phys. Lett.* **78**, 489–491 (2001).

Smith, D. R. *et al.* "Composite Medium with Simultaneously Negative Permeability and Permittivity." *Phys. Rev. Lett.* **84**, 4184–4187 (2000).

Smolyaninov, Igor I., Y. Hung and C. Davis. "Magnifying Superlens in the Visible Frequency Range." *Science* **315**, 1699–1701 (2007).

Valentine J. *et al.* "An optical cloak made of dielectrics." *Nature Materials* **8**, 568–571(2009).

Veselago, V. G. "The Electrodynamics of Substances with Simultaneously Negative Values of ε and μ." *Soviet Physics Uspekhi.* **10**, 509–514 (1968).

Wood, Ben. "Metamaterials and invisibility." *C. R. Physique* **10**, 379–390 (2009).

Yoshida, Takumi *et al.* "Transparent Cockpit," in *Proc. IEEE Conference on Virtual Reality, 8–12 March 2008, Reno, Nevada* (2008). pp. 185–188.

Zhang, Baile, Tucker Chan, and Bae-Ian Wu. "Lateral Shift Makes a Ground-Plane Cloak Detectable." *Phys. Rev. Lett.* **104**, 233903 (2010).

Zhang, Baile *et al.*" Macroscopic Invisible Cloak for Visible Light." ArXiv.org. http://arxiv.org/abs/1012.2238. Dec. 10, 2010, revised, Dec. 21, 2010. Accessed 12/28/10.

Chapter 7 Light Fantasy to Light Reality

Background and popular treatments

Adler, Robert. "Acoustic 'superlens' could mean finer ultrasound scans." *New Scientist* http://www.newscientist.com/article/dn13156-acoustic-superlens-could-mean-finer-ultrasound-scans.html. January 2008. Accessed 10/15/2010.

Barras, Colin. "Invisibility cloaks could take sting out of tsunamis." *New Scientist,* http://www.newscientist.com/article/dn14829. 29 September 2008. Accessed 10/15/2010.

Bohan, Suzanne. "Livermore lab nears launch of fusion quest." *Contra Costa Times* (Walnut Creek, Calif.). http://www.physorg.com/news204484818.html. Sept 23, 2010. Accessed 10/2/2010.

The British Interplanetary Society. "Warp Drive, Faster Than Light." http://www.bis-spaceflight.com/sitesia.aspx/page/1539/l/en-gb. 15 November 2007. Accessed 9/29/2010. "Papers from the workshop '*Warp Drive,*

Faster Than Light'." JBIS Papers. n.d. http://www.bis-spaceflight.om/sitesia.aspx/page/360/id/1825/l/en. Accessed 9/29/2010.

Cramer, John G. "The Alcubierre Warp Drive." *Analog Science Fiction & Fact,* November 1996, Alternate View Column AV-81. http://www.npl.washington.edu/AV/altvw81.html. "The Micro-Warp Drive." *Analog Science Fiction & Fact,* February 2000. Alternate View Column AV-99. http://www.npl.washington.edu/AV/altvw99.html. Accessed 9/20/2010.

———. "Faster-than-Light Implications of Quantum Entanglement and Nonlocality," in Marc G. Millis and Eric W. Davis, eds., *Frontiers of Propulsion Science.* Reston, VA: American Inst. of Aeronautics & Astronautics, 2009. pp. 509–529.

Hemmer, P. and J. Wrachtrup. "Where Is My Quantum Computer?" *Science* **324**, 473–474 (2009).

Hughes, Richard *et al.* "A Quantum Information Science and Technology Roadmap, Part 1." Advanced Research and Development Activity (ARDA). Report LA-UR-04-1778. http://qist.lanl.gov/pdfs/qc_roadmap.pdf. April 2, 2004. Accessed 9/29/10.

Japan Aerospace Exploration Agency. "Small Solar Power Sail Demonstrator "IKAROS" Confirmation of Photon Acceleration." http://www.jaxa.jp/press/2010/07/20100709_ikaros_e.html. July 9, 2010. Accessed 10/27/2010.

Lawrence Livermore National Laboratory. "LIFE: Clean Energy from Nuclear Waste." https://lasers.llnl.gov/about/missions/energy_for_the_future/life/. n.d. Accessed 10/1/2010.

———. "NIF: In the News — 2010." https://lasers.llnl.gov/newsroom/in_the_news/. Dec. 17, 2010. Accessed 12/26/2010.

Malik, Tariq. "Japanese Solar Sail Successfully Rides Sunlight." *Space.com.* http://www.space.com/businesstechnology/solar-sail-successfully-flies-on-sunlight-100712.html. 12 July 2010. Accessed 9/21/2010.

Matson, John. "Record 232-digit number from cryptography challenge factored." *Scientific American.* http://www.scientificamerican.com/blog/post.cfm?id=record-232-digit-number-from-crypto-2010-01-08. Jan. 8, 2010. Accessed 9/25/2010.

Marvin, Carolyn. *When Old Technologies Were New.* New York: Oxford University Press, 1988.

Nelson, Amy. "Technologies for research, energy are advanced at SPIE Optics and Photonics." *SPIE.* http://spie.org/x41494.xml. Aug. 9, 2010. Accessed 10/1/2010.

Optics.org. "Laser fusion energy: commercial by 2030?" http://optics.org/news/1/3/1. Aug. 2, 2010. Accessed 10/1/2010.

Petit, Charles. "Quantum Computer Simulates Hydrogen Molecule Just Right." *Science News* Jan 28, 2010. http://www.wired.com/wiredscience/2010/01/quantum-computer-hydrogen-simulation/. Accessed 9/25/2010.

Seaver, Lynda. "National Ignition Facility achieves unprecedented 1 megajoule laser shot." *Lawrence Livermore National Laboratory.* https://publicaffairs.llnl.gov/news/news_releases/2010/nnsa/NR-NNSA-10-01-02.html. Jan 27, 2010. Accessed 10/1/2010.

For readers who want more detail

Ahmed, Md. Farid *et al.* "A Review of One-Way and Two-Way Experiments to Test the Isotropy of the Speed of Light." http://arxiv.org/abs/1011.1318v1.

Brown, Kenneth R. "Quantum computing." *Nature Chemistry* **2**, 76–77 (2010).

Casimir, H. B. G. "On the attraction between two perfectly conducting plates." *Proc. Kon. Nederl. Akad. Wet.* **51**, 793 (1948).

Chaturvedi, Pratik *et al.* "A smooth optical superlens." *Appl. Phys. Lett.* **96**, 43102 (2010).

DiCarlo, L. *et al.* "Demonstration of two-qubit algorithms with a superconducting quantum processor." *Nature* **460**, 240–244 (2009).

Glenzer, S. H. *et al.* "Symmetric Inertial Confinement Fusion Implosions at Ultra-High Laser Energies." *Science* **327**, 1228 (2010).

Guenneau, Sébastien *et al.* "Acoustic metamaterials for sound focusing and confinement." *New J. Phys.* **9**, 399 (2007). http://iopscience.iop.org/1367-2630/9/11/399/fulltext. Accessed 10/15/2010.

Klaers, Jan *et al.* "Bose–Einstein condensation of photons in an optical microcavity." *Nature* **468**, 545–548 (2010).

Lanyon, B. P. *et al.* "Towards quantum chemistry on a quantum computer." *Nature Chemistry* **2**, 106–111 (2010).

Li, C. K. "Charged-Particle Probing of X-ray–Driven Inertial-Fusion Implosions." *Science* **327**, 1231 (2010).

Norreys, Peter A. "Controlling Implosion Symmetry Around a Deuterium-Tritium target." *Science* **327**, 1208–1210 (2010).

O'Brien, Jeremy L. "Optical Quantum Computing." *Science* **318**, 1567–1570 (2007).

R&D. "NNSA and LLNL announce first successful integrated experiment at NIF." http://www.rdmag.com/News/2010/10/General-Science-Photonics-NNSA-And-LLNL-Announce-First-Successful-Integrated-Experiment-At-NIF/. 10/7/2010. Accessed 10/13/2010.

Service, Robert F. "Diamond Feats Give Quantum Computing a Solid Boost." *Science* **329**, 616–617 (2010).

Small, Aaron C., James H. Johnston, and Noel Clark. "Inkjet Printing of Water 'Soluble' Doped ZnS Quantum Dots." *Eur. J. Inorg. Chem.* **2010**, 242–247 (2010).

Van Den Broeck, C. "A 'warp drive' with more reasonable total energy requirements." *Class. Quantum Grav.* **16**, 3973 (1999).

Zambonelli, Franco and Marco Mamei. "The Cloak of Invisibility: Challenges and Applications," in *IEEE Pervasive Computing* **1**, Piscataway, NJ: IEEE CS Press, 2002. pp. 62–70.

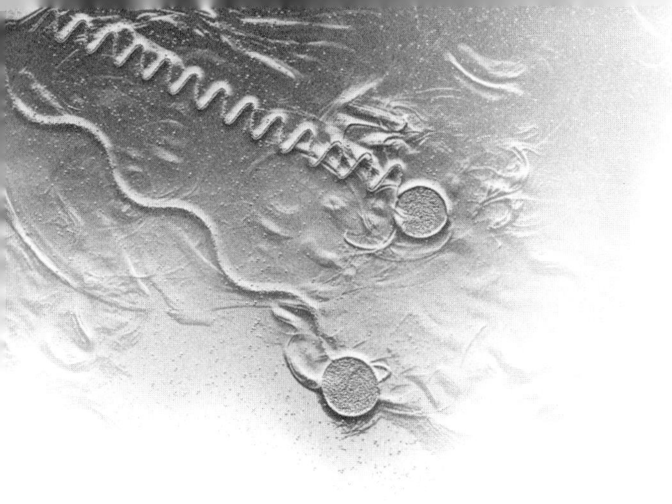

Index